DETRITUS AND DECOMPOSITION
IN
ECOSYSTEMS

NIPA® GENX ELECTRONIC RESOURCES & SOLUTIONS P. LTD.
New Delhi-110 034

www.nipaers.com
Browse, Search, Read & Buy...
Print Books
eChapters
eBooks
Publishers
Forthcoming
Subject Catalogues
New Titles
Online Resources on:
✓ Current Affairs
✓ Reasoning, Logic & Aptitude
✓ Competitive Examinations
✓ English Language Lab
✓ Writing & Pronunciation Tools
✓ Personality Development
✓ Online Programmes for Professionals
✓ Interview Preparation
Pay Using
PhonePe
Google Pay
Paytm
PayPal
UPI
CC Avenue

DETRITUS AND DECOMPOSITION IN ECOSYSTEMS

Authors

ZAFAR RESHI
Reader
Department of Botany, University of Kashmir
Srinagar-190 006, J&K, India

and

SUMIRA TYUB
Centre of Research for Development,
University of Kashmir
Srinagar-190 006, **J&K**, India

NIPA® GENX ELECTRONIC RESOURCES & SOLUTIONS P. LTD.
New Delhi-110 034

NIPA® GENX ELECTRONIC
RESOURCES & SOLUTIONS P. LTD.

101,103, Vikas Surya Plaza, CU Block
L.S.C. Market, Pitam Pura, New Delhi-110 034
Ph : +91 11 27341616, 27341717, 27341718
E-mail: newindiapublishingagency@gmail.com
web: www.nipabooks.com

For customer assistance, please contact
Phone: + 91-11-27 34 17 17
Fax: + 91-11- 27 34 16 16
E-Mail: feedbacks@nipabooks.com

ISBN: 978-81-19072-80-4

Composed and Designed by NIPA®.

Preface

Most of the Net Primary Production (annual gain of energy and matter by the plant sub-system) is not consumed by herbivores, but is returned to the environment as detritus which plays critical roles in organizing and sustaining ecosystems. In addition to its effects on trophic structure and dynamics, detritus physically alters habitats thereby facilitating some species and inhibiting others. It goes through a series of changes with time that are mediated by decomposers through the involvement of extracellular enzymes whose synthesis is affected by all the factors that affect microorganisms in soil, such as moisture, temperature, and available nutrients. After release, the activity of soil enzymes is regulated primarily by litter chemistry, substrate availability and temperature. Through their contribution to the recycling of essential chemical elements such as nitrogen and phosphorus (that are in limited supply for primary producers), decomposers play a key role in the primary productivity of ecosystems. While decomposition integrates collective activities of organisms within a soil food web, it has the potential to serve as an indicator of soil conditions also. Though the effect of detritus and decomposition on the structure and function of ecosystems has long been recognised, yet few attempts have been made to collate the existing data on various facets of this vital ecological process (detritus and its decomposition); a necessary pre-requisite for identifying broad patterns at local, regional and global scales. The present book attempts to bring together the enormous but fragmented information and fill this lacuna. It has been designed to outline the basic and fundamental aspects of detritus and its decomposition to be understood in their right perspective and envisages to put forward a clear understanding of the current concepts of this fundamental aspect of ecology and its widening horizons. The book, it is hoped, would serve the needs of graduate and postgraduate students and researchers interested in ecology and environmental sciences, soil biology and biochemistry, forestry, and other allied disciplines. Perfection is an extremely rare achievement and takes long time to be attained and it was not affordable for us at the present stage of inquisitiveness to wait that long; hence publication of the book. Suggestions and critical comments for improvements in future editions of the book will be highly appreciated.

The fruition of this book is the result of wealth of advice and feedback that the authors have received from their teachers, colleagues and students. In particular, the authors are indebted to Prof. B. A. Wafai, Department of Botany, University of Kashmir, Srinagar, for his stimulating and inspiring ideas and also for meticulous editing of the entire manuscript. Zafar Reshi is indebted to Professors B. L. Sapru and P. Kachroo, for their prescience in introducing him to the subject of Ecology and also for their encouragement during the early stages of his research career. Help extended by Professors G. H. Dar and S. Farooq in compiling this book is immensely acknowledged. Zafar Reshi also wishes to express deep sense of gratitude to his mother, wife and daughter for their patience in sparing several weekends and evenings which kept him occupied exclusively with this write up. Sumira Tyub is indebted to her parents and sister for the help and support provided by them during the compilation of this book.

The authors specially thank the entire staff of New India Pulishing Agency, New Delhi for painstaking efforts in bringing out the book in the present form.

July 2006
Srinagar.

Zafar Reshi
Sumira Tyub

CONTENTS

CHAPTER 1

NATURE AND COMPOSITION OF DETRITUS

1.1. Characteristics and categories

The term '**detritus**' is broadly defined as any form of non-living organic matter, including different types of plant tissues (e.g. leaf litter, dead wood, aquatic macrophytes, algae), animal tissues (carrion), dead microbes, faeces (manure, dung, faecal pellets, guano, frass), as well as products secreted, excreted or exuded by organisms (e.g. extracellular polymers, nectar, root exudates and leachates, dissolved organic matter, extracellular matrix, mucilage)(Moore *et al.*, 2004).

Chemical composition of detritus reflects its source, legacy and the biochemical composition. The compounds comprising detritus include, specialized polymers associated with cell walls of plants (cellulose, hemicellulose, lignin) and fungi (chitin, tannin and melanin) as well as universal biomolecules, such as fats, oils, nucleic acids, proteins and other polysaccharides and the monomeric constituents of these polymers like fatty acids, sugars, amino acids, nucleotides and nucleosides and other aliphatics and aromatics.

Size classifications of detritus range from simple and complex organic molecules in dissolved organic matter (DOM) to particulate organic matter (POM), plant litter and coarse woody debris. The size classification is, however, not standardized as the fractions are dynamic and interconvertible. For example, coarse debris is converted to fine particulate matter and dissolved forms in detritus are continually leached from all size fractions. Besides, dissolved detritus can be resynthesized into particulate fractions with small particulates building up into larger complexes upon aggregation.

Ecologists have routinely included detritus in the description of communities and ecosystems and have also studied its fate (Swift *et al.*, 1979). However, much of the descriptive work is confined to litter.

1.2. Litter

It refers to the layer of dead plant material that may be present on the soil surface or the dead plant material that is not attached to a living plant. These definitions, however, are not quite satisfactory for the ecologists as the litter layer may either be clearly distinguishable from an underlying mineral layer or there may be no sharp boundary between a layer containing recognizable plant structures and a layer containing only amorphous organic material. International Biological Programme (IBP) Hand book, "*Methods of Study on Quantitative Ecology*", defines litter as the material lying on the soil surface composed of dead plants and shed organs, but not standing dead matter (Medwecka-Kornas, 1971). Rodin and Bazilevich (1967) in a more extensive description included all dead organic matter from above- and below-ground parts; whether they die annually or whether they are added to the litter as a result of slower ageing process or natural thinning. In both cases, standing dead material is not measured in estimates of litter production (Bell, 1974). Rodin and Bazilevich (1967) included flowers, leaves, glumes, seeds, fruits and small twigs also in the leaf litter component while Facelli and Pickett (1991a) include only dead plant material of small size lying loose on ground excluding boles and large branches in the litter fraction. Medwecka—Kornas (1971), however, proposed inclusion of all material from the same community in the litter estimates, isolating neither true leaf material nor the "leaf fraction" as defined by Rodin and Bazilevich (1967).

Patterns of accumulation

Amount of litter accumulated in a given site is determined by the balance among various factors, like *in situ* litter production, deposition of litter from outside the system, litter destruction by physical and biotic agents and removal of litter. Litterfall represents an essential link in the organic production-decomposition cycle and thus is a fundamental ecosystem process (Meentemeyer, *et al.,* 1982; Pragassan and Parthasarathy, 2005).

Litter mass data are mainly available for forest ecosystems although limited information is also available on savanna, grasslands and deserts and the same is compiled in Table 1.1.

Table 1.1. Leaf and total litterfall in different stands and ecosystems

Forest	*Location*	*Leaf litterfall* ($g\ m^{-2}\ yr^{-1}$)	*Total litterfall* ($g\ m^{-2}\ yr^{-1}$)	*Reference*
Coniferous stands				
Abies firma	Japan	313	634	Ando, 1977
Abies firma & Tsuga sieboldii	Japan	299	462	Furuno, 1979
Abies firma & Tsuga sieboldii	Japan	277	509	Furuno, 1979
Abies sachalinensis	Japan	273	352	Hardiwinoto *et al.*, 1991
Agathis dammara	Java	350	620	Bruijnzeel, 1985
Agathis dammara	Java	270	580	Bruijnzeel, 1985
Chamaecyparis obtusa	Japan	279	415.6	Ueda & Tsutsumi, 1979
Chamaecyparis obtusa	Japan	216	349	Ueda & Tsutsumi, 1979
Chamaecyparis obtusa	Japan	325	484	Ueda & Tsutsumi, 1979
Chamaecyparis obtusa	Japan	379	482	Ueda & Tsutsumi, 1979
Chamaecyparis obtusa	Japan	355	469	Ueda & Tsutsumi, 1979
Chamaecyparis obtusa	Japan	276	390	Ueda & Tsutsumi, 1979
Chamaecyparis obtusa	Japan	372	470	Tsutsumi *et al.*, 1983
Chamaecyparis obtusa	Japan	254	313	Tsutsumi *et al*, 1983
Chamaecyparis obtusa	Japan	257	306	Tsutsumi *et al.*, 1983
Chamaecyparis obtusa	Japan	265	301	Tsutsumi *et al*, 1983
Cunninghamia lanceolata	China	145	300	Wu *et al.*, 1990

Contd...

Cunninghamia lanceolata	China	256	537	Wen *et al.*, 1989
Cunninghamia lanceolata	China	203	453	Wen *et at*, 1989
Cunninghamia lanceolata	China	185	395	Wen *et al.*, 1989
Cunninghamia lanceolata	China	185	291	Nie, 1994
Cunninghamia lanceolata	China	216	359	Nie, 1994
Cunninghamia lanceolata	China	264	434	Nie, 1994
Cunninghamia lanceolata	China	94	167	Yu, 1994
Cunninghamia lanceolata	China	180	480	Lian & Zhang, 1998
Cunninghamia lanceolata	China	235	448	Tian, 1989
Larix gmelini	China	334	357	Liu *et al.*, 1990
Larix leptolepsis	Japan	235	356	Kawahara et *al.*, 1981
Larix leptolepsis	Japan	334	381	Kawahara et *al.*, 1981
Larix sibirica	Siberia	274	243	Vedrova, 1995
Larix sp.	China	274	207	Cheng, 1984
Picea sp.	Russia	219	432	Remezov *et al.*, 1959
Picea sp.	Russia	-	430	Remezov *et al.*, 1959
Picea sp.	Russia	-	339	Remezov *et al.*, 1959
Picea sp.	Russia	-	510	Remezov *et al.*. 1959
Picea sp.	Russia	390	680	Remezov *et al.*, 1959
Picea sp.	Murmansk Province	-	300	Manakov, 1962
Picea sp.	Murmansk Province	-	180	Manakov, 1962

Contd...

Picea & Abies spp.	China	-	167	Cheng, 1984
Picea abies	Germany	-	300	Ehwald, 1957
Picea abies	Germany	-	300	Ehwald, 1957
Picea abies	Germany	-	270	Ehwald, 1957
Picea abies	Sweden	330	470	Nihlgard, 1972
Picea abies	Germany	-	200	Ehwald, 1957
Picea abies	Norway	199	328	Rosberg & Stuanes, 1992
Picea abies	Norway	143	229	Rosberg & Stuanes, 1992
Picea abies	Finland	185	283	Viro, 1955
Picea abies	Finland	187	240	Viro, 1955
Picea abies	Finland	174	216	Viro, 1955
Picea abies	Russia	143	209	Kazimirov & Morozova,1973
Picea abies	Russia	231	351	Kazimirov & Morozova,1973
Picea abies	Russia	251	384	Kazimirov & Morozova,1973
Picea abies	Russia	273	423	Kazimirov & Morozova,1973
Picea abies	Russia	181	282	Kazimirov & Morozova,1973
Picea abies	Russia	151	231	Kazimirov & Morozova,1973
Picea abies	Russia	285	451	Kazimirov & Morozova,1973
Picea abies	Russia	312	508	Kazimirov & Morozova,1973
Picea abies	Russia	140	198	Kazimirov & Morozova,1973

Contd...

Picea abies	Russia	261	415	Kazimirov & Morozova, 1973
Picea abies	Russia	220	330	Kazimirov & Morozova, 1973
Picea abies	Russia	286	439	Kazimirov & Morozova, 1973
Picea abies	Russia	288	441	Kazimirov & Morozova, 1973
Picea abies	Russia	263	395	Kazimirov & Morozova, 1973
Picea abies	Russia	248	362	Kazimirov & Morozova, 1973
Picea abies	Russia	225	328	Kazimirov & Morozova, 1973
Picea abies	Russia	217	318	Kazimirov & Morozova, 1973
Picea abovata	Siberia	73	94	Vedrova, 1995
Picea excelsa	Arkhan-gel Province	277	460	Marchenko & Karlov, 1962
Picea excelsa	Arkhan-gel Province	277	460	Marchenko & Karlov, 1962
Picea excelsa	Russia	271	360	Rudanova et *al.*, 1967
Picea excelsa	Russia	271	360	Rudanova et *al.*, 1967
Picea glehnii	Japan	258	369	Rudanova et *al.*, 1967
Picea sp.	China	265	336	Wang & Wang, 1991b
Picea sp.	China	292	369	Wang & Wang, 1991b

Contd...

Picea sp.	China	-	244	Cheng, 1984
Pinus halepensis	Mallorca	250	298	Garcia-Ple *et al.*, 1995
Pinus koraiensis	China	-	384	Cheng, 1984
Pinus koraiensis	China	217	320.7	Ding *et al.*, 1986
Pinus koraiensis	China	79	92	Ding *et al.*, 1986
Pinus koraiensis	China	89	122	Ding *et al.*, 1986
Pinus koraiensis	China	94	122	Ding *et al.*, 1986
Pinus koraiensis	China	107	162	Ding *et al.*, 1986
Pinus koraiensis	China	294	438	Wang & Wang, 1991a
Pinus koraiensis	China	253	362	Wang & Wang, 1991a
Pinus koraiensis	China	235	340	Wang & Wang, 1991a
Pinus massoni-ana	China	323	402	W u *et al.*, 1990
Pinus massoni-ana	China	384	572	Wen *et al.*, 1989
Pinus merkusii	Java, Indonesia	-	900	Gunadi, 1994
Pinus merkusii	Java, Indonesia	-	400	Gunadi, 1994
Pinus nigra	Europe	440	444	Berg *et al.*, 1993
Pinus pinaster	Europe	340	457	Berg *et al.*, 1993
Pinus pinaster	Spain	284	284	Moro *et al.*, 1996
Pinus pinaster	Spain	167	174	Hernandez *et al.*, 1992
Pinus pinaster	Italy	460	770	van Wesemael & Veer, 1992
Pinus pinea	Europe	501	618	Berg et *al.*, 1993

Contd...

Pinus pinea	Spain	213	240	Hernandez *et al.*, 1992
Pinus radiata	Europe	121	228	Berg *et al.*, 1993
Pinus radiata	Europe	266	317	Berg *et al.*, 1993
Pinus radiata	Europe	352	402	Berg *et al.*, 1993
Pinus roxburghii	Central Himalaya	257	426	Chaturvedi & Singh, 1987
Pinus roxburghii	Central Himalaya	425	664	Chaturvedi & Singh, 1987
Pinus roxburghii	Central Himalaya	432	662	Chaturvedi & Singh, 1987
Pinus roxburghii	Central Himalaya	553	738	Chaturvedi & Singh, 1987
Pinus roxburghii	Dehra Dun (India)	478	967	Pande & Sharma, 1986
Pinus roxburghii	Central Himalaya	-	865	Chaturvedi & Singh, 1987
Pinus sibirica	Siberia	322	348	Vedrova, 1995
Pinus sibirica	Japan	519	613	Furuno, 1993
Pinus sylvestris	Spain	363	713	Gallardo *et al.*, 1995
Pinus sylvestris	China	155	279	Shen *et al.*, 1996
Pinus sylvestris	China	103	259	Ding *et al.*, 1988
Pinus sylvestris	China	109	242	Ding *et al.*, 1988
Pinus sylvestris	China	256	499	Ding *et al.*, 1988
Pinus sylvestris	China	96	153	Ding *et al.*, 1988
Pinus sylvestris	Germany	-	300	Ehwald, 1957
Pinus sylvestris	Germany	-	240	Ehwald, 1957
Pinus sylvestris	Europe	140	238	Breymeyer, 1991
Pinus sylvestris	Europe	116	233	Breymeyer, 1991
Pinus sylvestris		-	402	de Visser *et al.*, 1994
Pinus sylvestris	Europe	125	210	Berg *et al.*, 1993
Pinus sylvestris	Europe	157	210	Berg *et al.*, 1993
Pinus sylvestris	Europe	142	303	Berg *et al.*, 1993

Contd...

Pinus sylvestris	Europe	186	328	Berg *et al.*, 1993
Pinus sylvestris	Europe	197	388	Berg *et al.*, 1993
Pinus sylvestris	Europe	200	369	Berg *et al.*, 1993
Pinus sylvestris	Europe	224	400	Berg *et al.*, 1993
Pinus sylvestris	Europe	94	162	Berg *et al.*, 1993
Pinus sylvestris	Europe	231	420	Berg *et al.*, 1993
Pinus sylvestris	Helsinki	111	166	Mälkönen, 1974
Pinus sylvestris	Helsinki	49	58	Mälkönen, 1974
Pinus sylvestris	Europe	69	83	Berg *et al.*, 1993
Pinus sylvestris	Europe	151	221	Berg *et al.*, 1993
Pinus sylvestris	Europe	73	156	Breymeyer, 1991
Pinus sylvestris	Finland	156	265	Viro, 1955
Pinus sylvestris	Finland	171	234	Viro, 1955
Pinus sylvestris	Finland	125	182	Viro, 1955
Pinus sylvestris	Finland	80	107	Mälkönen, 1974
Pinus sylvestris	Finland	219	285	Paavilainen, 1980
Pinus sylvestris	Finland	170	227	Paavilainen, 1980
Pinus sylvestris	Finland	170	227	Paavilainen, 1980
Pinus sylvestris	Finland	186	245	Paavilainen, 1980
Pinus sylvestris	Finland	183	245	Paavilainen, 1980
Pinus sylvestris	Finland	199	262	Paavilainen, 1980
Pinus sylvestris	Europe	148	202	Berg *et al.*, 1993
Pinus sylvestris	Siberia	260	294	Vedrova, 1995
Pinus sylvestris	Europe	71	136	Berg *et al.*, 1993
Pinus sylvestris	Europe	91	168	Berg *et al.*, 1993
Pinus sylvestris	Europe	69	208	Berg *et al.*, 1993
Pinus sylvestris	Europe	53	74	Berg *et al.*, 1993
Pinus sylvestris	Europe	61	92	Berg *et al.*, 1993
Pinus sylvestris	Europe	32	58	Breymeyer, 1991

Contd...

Pinus tabulaeformis	China	-	268	Yao, 1989
Pinus tabulaeformis & Platycladus orientalis	China	-	320	Yao, 1989
Pinus tabulae-formis	China	299	456	Nie, 1985
Pinus taeda	Japan	535	575	Akai & Furuno, 1970
Pinus taeda	Japan	638	680	Akai & Furuno, 1970
Pinus taeda	Japan	704	765	Akai & Furuno, 1970
Platycladus orientalis	China	-	397	Yao, 1989
Sciadopitys verticillata	Shiga (Japan)	350	613	Ohkubo *et al.*, 1981
Broad leaved stands				
Alnus nepalensis	Eastern Himalaya	416	520	Sharma & Ambasht, 1987
Alnus nepalensis	Eastern Himalaya	453	566	Sharma & Ambasht, 1987
Alnus nepalensis	Eastern Himalaya	453	566	Sharma & Ambasht, 1987
Alnus nepalensis	Eastern Himalaya	436	545	Sharma & Ambasht, 1987
Arundinaria falcata	Central Himalaya	174	351	Rawat *et al.*, 1994
Betula fruticosa	Siberia	197	206	Vedrova, 1995
Betula platyphylla var. japonica	Hokkaido (Japan)	219	300	Hardiwinoto *et al*, 1991
Betula platyphylla	China	300	343	Zhang & Zhou, 1991
Betula platyphylla	China	280	524	Bao, 1991
Betula platyphylla	Helsinki	188	235	Mölkönen, 1977
Betula sp.	Finland	127	181	Viro, 1955
Betula sp.	Finland	94	150	Viro, 1955

Contd...

Betula verrucosa	Russia	360	520	Smirnova & Gorodentseva, 1958
Bruguiera & Rhizophora spp.	Malaysia	-	1180	Ong *et al.*, 1980
Bruguiera gymnorrhiza	South Andaman	348	511	Dagar & Sharma, 1993
Bruguiera gymnorrhiza	South Andaman	520	709	Dagar & Sharma, 1993
Bruguiera sexangula	China	-	1104	Lu & Lin, 1990
Bruguiera sexangula	China	-	1255	Lu & Lin, 1990
Butea sp.	Varanasi (India)	84	101	Singh, 1968
Castanea sativa	Spain	324	625	Gallardo *et al.*, 1995
Castanopsis hystrix	China	520	759	Zheng *et al.*, 1995
Castanopsis kawakamii	Chinae	810	1300	Lian & Zhang, 1998
Casuarina equisetifolia	India	392	552	Hosur *et al.*, 1997
Cyclobalanopsis glauca	Zhejiang (China	-	478	Yu, 1996
Cyclobalanopsis spp.	Japan	356	488	Kira, 1978
Cyclobalanopsis spp.	Japan	457	589	Kira, 1978
Cyclobalanopsis spp.	Japan	422	554	Kira, 1978
Decussocarpus nagi	Japan	288	490	Watanabe, 1978
Dendrocalamus strictus	India	460	718	Tripathi & Singh, 1994
Dendrocalamus strictus	India	460	591	Tripathi & Singh, 1996
Dendrocalamus strictus	India	583	583	Hosur *et al.*,1997

Contd...

Diospyros-Anogeissus	Varanasi (India)	368	421	Singh, 1968
Dipterocarp	Malaysia	540	880	Proctor *et al*., 1983
Dipterocarp	Malaysia	670	1090	Proctor *et al*., 1983
Dipterocarp	Thailand		470	Thaiutsa *et al*., 1978
Englehardtia spicata	Cherra-punji (India)	910	1178	Khiewtam & Ramakrishnan, 1993
Erythrophloem fodii	China	421	638	Wu *et al*., 1990
Eucalyptus citriodora	China	348	515	Wen *et al*., 1989
Eucalyptus spp.	Dehradun (India)	591	707	Pande & Sharma, 1986
Eucalyptus tereticornis	India	604	820	Hosur *et al*., 1997
Fagus crenata	Ashiu (Japan)	345	483	Ogino, 1977
Fagus crenata	Japan	237	296	Kakubari, 1977
Fagus crenata	Japan	244	303	Kakubari, 1977
Fagus crenata	Japan	276	335	Kakubari, 1977
Fagus crenata	Japan	397	505	Kakubari, 1977
Fagus crenata	Japan	362	470	Kakubari, 1977
Fagus crenata	Japan	238	251	Kakubari, 1977
Fagus crenata	Japan	185	239	Kakubari, 1977
Fagus crenata	Japan	190	244	Kakubari, 1977
Fagus sylvatica	Germany		280	Ehwald, 1957
Fagus sylvatica	China	-	350	Ehwald, 1957
Fagus sylvatica	China	-	250	Ehwald, 1957
Fagus sylvatica	South Sweden	260	470	Nihlgard, 1972
Fagus sylvatica	South Sweden	260	470	Nihlgard, 1972
Fagus sylvatica	Sweden	260	470	Nihlgard, 1972

Contd...

Grewia tiliifolia	Kaumon (India)	1075	1420	Singh *et al.*, 1993
Ixomanthes chinensis	China	355	559	Chen *et al.*, 1992
Lantana camara	Central Himalaya	310	520	Rawat *et al.*, 1994
Leacaena leucocephala	Oga-sawara Islands	498	1229	Kimura *et al.*, 1984
Machilus thunbergii	Japan	529	748	Ueda & Tsutsumi, 1980
Machilus thunbergii	Japan	445	614	Ueda & Tsutsumi, 1980
Poeciloneuron indicum	Karnata-ka	288	344	Rai & Proctor, 1986
Populus tremula	Germany	-	410	Ehwald, 1957
Populus tremula	Germany	-	490	Ehwald, 1957
Populus tremula	Russia	350	550	Remezov *et al.*, 1959
Populus tremula	Russia	-	390	Remezov *et al.*, 1959
Populus tremula	Siberia	374	389	Vedrova, 1995
Quercus sp.	Kyoto	374	578	Furuno & Saito, 1981
Quercus sp.	China	-	310	Ehwald, 1957
Quercus acutissima	China	543	717	Wen *et al.*, 1989
Quercus cerris, Quercus petraea	Tuscany (Italy)	530	670	van Wesemael & Veer, 1992
Quercus flori-bunda	Central Himalaya	375	478	Rawat & Singh, 1988a
Quercus ilex, Quercus cerris	Tuscany (Italy)	290	420	van Wesemael & Veer, 1992
Quercus lanuginosa	Central Himalaya	590	781	Rawat & Singh, 1988a
Quercus leucotrichophora	Central Himalaya	461	580	Rawat & Singh, 1988b
Quercus leucotrichophora	Garhwal Himalaya	729	934	Pant & Tiwari, 1992

Contd...

Quercus mongolia	China	-	460	Wei, 1991
Quercus mongolia var. grossesrrata	Hokkaido (Japan)	251	343	Hardiwinoto *et al.*, 1991
Quercus petraea	Hungary	445	589	Jakucs, 1985
Quercus petraea	Scotland	236	361	Tavakol & Proctor, 1994a
Quercus petraea	Scotland	299	447	Tavakol & Proctor, 1994b
Quercus petraea	The Netherlands	367	671	Drift, 1974
Quercus petraea	Scotland	407	525	Tavakol & Proctor, 1994a
Quercus pyrenaica	Spain	363	562	Gallardo *et al.*, 1995
Quercus robur	Belgium	317	526	Duvigneaud, 1970
Quercus robur	Russia	356	496	Goryshina, 1974a; 1974b
Quercus robur	Russia	376	606	Goryshina, 1974a; 1974b
Quercus robur	Russia	370	570	Remezov *et al.*, 1959
Quercus rotundifolia	Spain	185	232	Hernandez *et al.*, 1992
Quercus sp.	Kyoto (Japan)	318	429	Furuno & Saito, 1981
Quercus sp.	China	-	330	Ehwald, 1957
Quercus sp.	China	-	430	Ehwald, 1957
Quercus sp.	China	-	520	Ehwald, 1957
Quercus sp.	China	-	410	Ehwald, 1957
Quercus sp.	China	-	410	Ehwald, 1957
Quercus variabilis	Jiangsu (China)	442	736	Yu, 1994
Quercus ilex, Quercus petraea	Tuscany (Italy)	250	400	van Wesemael & Veer, 1992
Quercus suber, Pinus pinaster	Tuscany (Italy)	80	260	van Wesemael & Veer, 1992

Contd...

Quercus suber, Quercus cerris	Tuscany (Italy)	490	650	van Wesemael & Veer, 1992
Rhizophora & Bruguiera spp.	Malaysia	-	1060	Ong *et al.*, 1980
Rhizophora & Bruguiera spp.	Malaysia	-	830	Ong *et al.*, 1980
Rhizophora & Bruguiera spp.	Malaysia	-	1420	Ong *et al.*, 1980
Rhizophora sp.	Malaysia	-	1490	Ong *et al.*, 1980
Rhododendron sp.	Central Himalaya	167	346	Garkoti & Singh, 1995
Schima superba	China	420	710	Lian & Zhang, 1998
Schima wallichii	China	327	434	Wu *et al.*, 1990
Shorea curtisii, Gluta elegans & Anisoptera curtisii	Malaysia	536	745	Gong & Ong, 1983
Shorea robusta	UP (India)	685	685	Singh *et al.*, 1993
Shorea robusta	Dehradun (India)	501	555	Pande & Sharma, 1986
Shorea-Buchanania spp.	Varanasi (India)	302	315	Singh, 1968
Tectona sp.	Varanasi (India)	474	502	Singh, 1968
Tectona grandis	UP (India)	770	770	Singh *et al.*, 1993
Tectona grandis	Dehradun (India)	829	1127	Pande & Sharma, 1986
Tilia cordata syn. parvifolia	Russia	290	450	Remezov *et al.*, 1959
Tilia cordata syn. parvifolia	Russia	450	520	Remezov *et al.*, 1959
Trerminalia shorea	Varanasi (India)	609	621	Singh, 1968
Xylia xylocarpa	Western Ghats (India)	1069	1444	Kumar & Deepu, 1992

Contd...

Xylia xylocarpa	Western Ghats (India)	816	1218	Kumar & Deepu, 1992
Xylia xylocarpa	Western Ghats (India)	875	1346	Kumar & Deepu, 1992
Ecosystem types				
Alluvial forest	Malaysia	660	1150	Proctor *et al*., 1983
Dry deciduous forest	Varanasi (India)	-	170	Pandey *et al*., 1980
Dry deciduous forest	Varanasi (India)	150	210	Sharma, 1981
Dry deciduous forest	Varanasi (India)	-	400	Garg & Vyas, 1975
Dry deciduous forest	Varanasi (India)	130	150	Gaur & Pandey, 1978
Dry deciduous forest	Varanasi (India)	170	180	Gaur & Pandey, 1978
Dry deciduous forest	Varanasi (India)	420	570	Singh & Misra, 1978
Dry deciduous forest	Varanasi (India)	390	510	Singh & Misra, 1978
Dry deciduous forest	Varanasi (India)	-	780	Singh, 1975
Dry deciduous forest	Varanasi (India)	500	-	Singh, 1968
Dry deciduous forest	Varanasi (India)	420	-	Singh, 1968
Dry deciduous forest	Varanasi (India)	480	-	Ranawat & Vyas, 1975
Dry deciduous forest	Varanasi (India)	620	-	Singh, 1968
Dry formation	West Malaysia	626	787	Crowther, 1987
Dry formation	West Malaysia	438	605	Crowther, 1987
Dry formation	West Malaysia	377	519	Crowther, 1987

Contd...

Dry formation	West Malaysia	449	641	Crowther, 1987
Dry formation	West Malaysia	404	573	Crowther, 1987
Dry formation	West Malaysia	894	1205	Crowther, 1987
Dry formation	India	290	380	Balasubramanyan,1964
Dry formation	India	620	770	Bandhu, 1973
Dry sub-humid to humid grasslands	India	-	1-433	Melkania & Singh 1989
Dry tropical forest	India	-	488-671	Singh, 1992
Evergreen forest	Karnataka	290	340	Rai, 1981
Evergreen forest	Karnataka	340	400	Rai, 1981
Evergreen forest	Karnataka	320	410	Rai, 1981
Evergreen forest	Kerala	-	440-640	Rajendraprasad *et al.*, 2000
Evergreen forest	Karnataka	360	420	Rai, 1981
Evergreen forest	Hong Kong	836	1219	Lam & Dudgeon, 1985
Evergreen forest	China	451	799	Wen *et al.*, 1989
Evergreen forest	Guangdong (China)	530	820	Tu, 1984
Forest over limestone	Malaysia	730	1200	Proctor *et al.*, 1983
Fresh water swamp	Malaysia	720	920	Furtado *et al*, 1980
Heath forest	Malaysia	560	920	Proctor *et al.*, 1983
High altitude forest	Central Himalaya	170-370	350-630	Garkoti & Singh, 1995
High elevational Himalayan grasslands	India	-	15-376	Melkania & Singh 1989
Hill evergreen	Thailand	-	690	Thaiutsa *et al*, 1978

Contd...

Low land evergreen	Ceylon	-	640	Hladik & Hladik, 1972
Low-elevational Himalayan grasslands	India	-	3-232	Melkania & Singh 1989
Lower montane forest	Java	540	680	Bruijnzeel, 1982
Lowland dipterocarp	Malaysia	640	890	Lim, 1978
Lowland dipterocarp	Malaysia	630	1060	Ogawa, 1978
Lowland dipterocarp	Malaysia	540	750	Gong & Ong, 1983
Lowland dipterocarp	Tropical Mindanao	-	530	Kellman, 1970
Mangroves	Thailand	-	930	Aksornkoae & Khemnark, 1980
Mangroves	Malaysia	-	1000	Gong *et al.*, 1983
Mangroves	Malaysia	-	1030	Gong *et al.*, 1983
Mangroves	Malaysia	-	1140	Gong *et al.*, 1983
Mangroves	Malaysia	-	760	Gong *et al.*, 1983
Mangroves	Japan	-	590	Nishira, 1983
Mangroves	Japan	-	450	Nishira, 1983
Mangroves	Japan	-	280	Nishira, 1983
Mangroves	Japan	-	300	Nishira, 1983
Mid-elevational Himalayan grasslands	India	-	2-80	Melkania & Singh 1989
Moist deciduous forest	Western Ghats	900-1160	1300-1500	Swamy & Proctor, 1994
Montane evergreen forest	Gabon	350	590	Hladik, 1978
Montane forest	Java	450	600	Yamada, 1976

Contd...

Montane forest	India	390	560	Blasco & Tassy, 1975.
Montane rain forests	China	542	771	Lu & Liu, 1994
Montane rain forests	China	745	975	Lu & Liu, 1994
Rain forest	Karnata-ka	360	418	Rai & Proctor, 1986
Rain forest	Karnata-ka	342	398	Rai & Proctor, 1986
Rain forest	Karnata-ka	322	407	Rai & Proctor, 1986
Rain forests	Sumatra	762	1114	van Schaik & Mirmanto, 1985
Rain forests	Sumatra	831	1137	van Schaik & Mirmanto, 1985
Rain forests	Sumatra	738	1017	van Schaik & Mirmanto, 1985
Rain forests	Sumatra	632	987	van Schaik & Mirmanto, 1985
Rain forests	Sumatra	734	1103	van Schaik & Mirmanto, 1985
Rain forests	Sumatra	608	961	van Schaik & Mirmanto, 1985
Rain forests	Sumatra	564	827	van Schaik & Mirmanto, 1985
Rain forests	Sumatra	559	835	van Schaik & Mirmanto, 1985
Rain forests	Sumatra	669	1155	van Schaik & Mirmanto, 1985
Rain forests	Sumatra	597	1024	van Schaik & Mirmanto, 1985
Secondary forest	Tropical Mindanao	-	1250	Kellman, 1970
Secondary forest	Tropical Mindanao	-	760	Kellman, 1970
Secondary forest	Tropical Mindanao	-	1110	Kellman, 1970
Secondary forest	Tropical Mindanao	-	1220	Kellman, 1970

Contd...

Secondary forest	Tropical Mindanao	-	1000	Kellman, 1970
Secondary forest	Tropical Mindanao	-	700	Kellman, 1970
Secondary forest	Tropical Mindanao	-	940	Kellman, 1970
Semi-arid to arid grasslands	India	-	6-517	Melkania & Singh, 1989
Semi-dry deciduous forest	Varanasi	-	390	Gaur & Pandey, 1978
Semi-dry deciduous forest	Varanasi	240	460	Gaur & Pandey, 1978
Subtropical forest	Megha-laya (India)	470	650	Singh & Ramakrishnan, 1982
Subtropical forest	Megha-laya (India)	380	650	Singh & Ramakrishnan, 1982
Teak forest	Thailand		790	Thaiutsa *et al.*, 1978
Tropical dry evergreen forest	South India	900-960	1330-1350	Pragasan & Parthasarathy, 2005
Tropical dry evergreen forest	Coroman-del coast	340-950	510-1110	Visalakshi, 1993
Dry-mixed deciduous forest	Varanasi	320	430	Rai & Srivastava, 1982
Tropical ever-green forest	Western Ghats	780-1170	970-1420	Swamy & Proctor, 1994
Tropical forest	Costa Rica	-	530-900	Heaney & Proctor, 1989
Tropical forest	Amazonia	210-760	240-1030	Cuevas & Medina, 1986
Tropical rain forest	Australia	520-620	800-1100	Stocker *et al.* 1995
Tropical rain forest	North Queens-land	350-430	500-600	Herbohn & Congdon, 1993

Contd...

Tropical rainforest	Colombia	280-460	430-700	Veneklaas, 1991
Tropical semievergreen forest	Western Ghats	-	673	Parthasarathy, 1992
Tropical wet evergreen forest	Western Ghats	360-490	614	Parthasarathy, 1992
Upper montane forest	Australia	570	1100	Proctor et *al.*, 1983.
Upper montane forest	Malaysia	230	360	Proctor et *al.*, 1983.

Source: Melkania & Singh (1989), Liu *et al.* (2004), Pragasan & Parthasarathy (2005), -data not available

Accumulation of litter occurs particularly in biomes (Table 1.2), such as scrub tundra, taiga and temperate forests (Ajtay *et al.*, 1979). Contrastingly, in wet tropical forests where mean temperature is above 30°C, litter decomposes faster, hence does not accumulate while as in areas with mean temperature range between 25°C to 30°C, litter supply and decomposition are almost equal.

Table 1.2. Estimated litterfall and litter in various ecosystem types.

Ecosystem type	**Area** ($m^2 \times 10^{12}$)	**Litterfall** ($g\ m^{-2}\ yr^{-1}$)	**Total litterfall** ($g \times 10^{15}$)
Forests	31.3		36.3
Tropical rain	10	1850	18.5
Tropical seasonal	4.5	1300	5.9
Mangrove	0.3	600	0.2
Temperate	6	850	5.1
Boreal (closed)	6.5	600	3.9
Boreal (open)	2.5	550	1.4
Forest plantations	1.5	875	1.3
Temperate woodlands	2	1220	2.4
Chaparral, bushland	2.5	1000	2.5
Savanna	22.5		31.3
Grass dominated	19	1500	28.5
Savanna forest	3.5	800	2.8

Contd...

Temperate grassland	12.5		8.6
Wet grassland	5	900	4.5
Dry grassland	7.5	550	4.1
Arctic/ alpine tundra	9.5		1.43
Polar desert	1.5	20	0.03
Herb-lichen tundra	3.6	145	0.52
Scrub tundra	4.4	200	0.88
Desert and semi-desert	21	125	2.6
Extreme desert	9	15	0.14
Bog, swamp and marshes	3.5	600	600

Source: Ajtay *et al.* (1979)

Notwithstanding limited data available on litter fall in different ecosystems (except forests) few attempts have been made to determine, with precision, geographic patterns of litter fall or its quantitative relationship to climatic variables. One of the earliest such attempts was made by Meentemeyer *et al.* (1982) who developed models that relate leaf litter production to appropriate climatic variables over the full range of terrestrial climates found on the Earth. Using computerized world climatic data base and mapping system (Box, 1975; 1978), first generation world maps of estimated plant litter production were obtained and subsequently quantified by computer planimetry to provide estimated world totals of leaf litter production (35.1 X 10^9 tons) and total litter production (54.8 x 10^9 tons). Computer assisted bivariate scatter diagrams (Meentemeyer *et al.*, 1982) reveal strong correlation between litter production and certain aspects of climate, viz., annual actual evaporation ($r = 0.89$), latitude ($r = 0.83$) and annual potential evaporation ($r = 0.64$). Stepwise multiple correlation-regression programme was used to develop a regression model that explains the largest amount of variance in total litter fall. Inclusion of latitude and potential evapotranspiration, however, did not significantly improve coefficient of determination (R^2). The resulting predictive equations developed by Meentemeyer *et al.* (1982) are reproduced below:

Total litterfall = 1.066 (AET)-183.98
Total litter fall = 433.6 – 8.185 (Latitude) + 0.5309 (AET)
Total litterfall = 888.59 – 11.35 (Latitude) – 3.63 (PAR) + 0.43 (AET)

AET = actual evapotranspiration; PAR = photosynthetically active radiation

Thus, actual evapotranspiration (AET) is a robust predictor of litterfall as it expresses the combined effect of temperature and water on ecological processes at a regional or global scale as was also reported by Berg and Meentemeyer (2001). In the latter study, litterfall data of 64 sites in Europe (mostly in Fennoscandia) comprising 48 sites of *Pinus spp.* (mostly *Pinus sylvestris*) and 16 of *Picea* spp. (mostly *Picea abies*) were used to study relationship between litterfall and actual evapotranspiration. The results of the study are summarized in Table 1.3.

Table 1.3. Relationship of litterfall in pine and spruce forests with actual evapotranspiration.

Forest	Intercept	Coefficient	r	R^2	n	p
Pine	-5651.08	15.785	0.766	0.587	48	<0.001
Spruce	-10129.3	25.539	0.891	0.794	16	<0.001

Adapted from Berg and Meentemeyer (2001).

Recently Liu *et al.* (2004) studied relationship between climatic factors and litterfall in coniferous and broadleaf forests in Eurasia. The study reveals that total litterfall in broadleaf forests of temperate, subtropical, and tropical areas is higher than in coniferous forests, but in the boreal zone opposite is true. At continental scale, total litterfall ranges from 30 to 500 g m^{-2} $years^{-1}$ in the boreal coniferous forests; to about 900 g m^{-2} $years^{-1}$ in tropical coniferous forests and about 1,500 g m^{-2} $years^{-1}$ in tropical broadleaf forests (Table 1.4). Litterfall is reported to increase significantly with annual mean temperature (T) and precipitation (P) in broadleaf forests than in coniferous forests. In both these forest types temperature has more pronounced effect than precipitation (Table 1.4)

Table 1.4. Relationship between total litterfall (Ln (L_{total}), (g m^{-2} $year^{-1}$) with mean annual temperature (Ln (T^{+}) °C) and annual precipitation (Ln (P) mm).

Forest	Constant Intercept	Ln(T^{+}) Coefficient	Ln (P) Coefficient	n	R^2_{adj}
Ln (L_{total}) = c + a x Ln (T^{+})					
Broad leaved	3.284	0.930		240	0.484
Coniferous	3.732	0.715		199	0.240
Both	3.120	0.962		439	0.516
Ln (L_{total}) = c + a x Ln (P)					
Broad leaved	2.295		0.557	240	0.376
Coniferous	2.668		0.454	199	0.199
Both	1.620		0.633	439	0.408
Ln (L_{total}) = c + a x Ln (T^{+}) + Bx Ln(P)					
Broad leaved	2.643	0.726	0.181	240	0.498
Coniferous	2.708	0.505	0.240	199	0.272
Both	2.296	0.741	0.214	439	0.535

Source: Liu *et al* . (2004). T^{+} = T + 10°C

Apart from climate, tree species diversity also exhibits close correlation with forest litter pools. In a regional study, covering 1,000 forests differing in tree species diversity in Catalonia (NE Spain), tree species richness, the identity of tree species and functional types showed significant influence on litter pools (Vilà *et al.*, 2004). General linear model of environmental and biotic effects on litter pools in forests of Catalonia is given in the following box:

Also herbivores may either increase or decrease the production of litter. As standing biomass is reduced by consumption, there is also decrease in the amount of litter produced (Facelli and Pickett, 1991a). The proportion of biomass consumed also varies considerably among systems; being negligible in young oldfields and mature woodlots where most of the biomass produced is transformed into litter (Odum, 1960). According to Whittaker and Woodwell (1969) only 9% of the yearly production in an oldfield is consumed by arthropods. Similar values are characteristic of temperate deciduous forests, and even lower (<2%)

Source	F ratio	p
Tree richness	13.82	0.0002
Tree functional type	2.548	0.08
Leaf area index	10.73	0.001
Productivity	5.08	0.02
Total basal area	12.21	0.0005
Shrub cover	0.18	0.67
Age	0.02	0.89
Evapotranspiration	1.25	0.61
Precipitation	0.96	0.33
Mean temperature	27.48	<0.0001
Hill site position	1.18	0.31
Seasonality	1.12	0.34

for evergreen forests (Waring and Schlesinger, 1985). In contrast, primary consumers constitute the main pathway of the annual production in grasslands, where large grazers substantially reduce the rates of litter accumulation (Weaver and Rowland, 1952; Dix, 1960; Hunt, 1978; Knapp and Seastedt, 1986). Herbivores sometimes increase the production of litter temporarily because of the damage to and subsequent death of unconsumed organs (Owen, 1978; Hollinger, 1986; Choudhury, 1988; Risley and Crossley, 1988). Some herbivores like grasshoppers cut more leaves than they consume (Rodell, 1978) and hence cause increase in litter production.

Existence of time lags between the formation of a plant organ and its deposition as litter is one of the factors that is largely relevant at small scale, though it is unimportant at large temporal and spatial scales (Facelli and Pickett, 1991a). Time lags depend on the dominant life form (Golley, 1965). Far less than 50% of the annual production of woodlands becomes litter within a year (Olson, 1963), but in herb dominated oldfields, most aerial biomass is transformed into litter at the end of the growing season (Golley, 1965). Differences in time lags may be important even within biomes as standing dead material is transformed faster into litter in semi-arid and moist grasslands than in sub-humid ones (Hunt, 1978). Trampling, snow packing, winds and storms affect the speed of this transformation. In addition, different organs have different characteristic time lags. Dead branches usually remain on shrubs and trees for a long time, but most leaves fall as they die (Olson, 1963; Sprugel, 1984; Noy-Meir, 1985). Ina forest undergoing wave regeneration, fall of leaves as litter occurs immediately after death of the whole canopy, but fall of dead branches and bark is delayed for several years (Sprugel, 1984).

Deposition of exogenous litter and the removal of the native litter are seldom considered in litter balances (production and disappearance of litter). The importance of litter movement depends mainly on the nature of transporting medium i.e., water or air (Polunin, 1984), the geometry and specific weight of the litter (Orndorff and Lang, 1981) and the geometry of the environment (Orndorff and Lang, 1981; MacMahon and Wagner, 1985; Noy-Meir, 1985). Litter movement is very important in aquatic systems where the energy transmitted by water is large, and litter floats easily (Polunin, 1984; Kenworthy *et al.*, 1987) and in deserts which have open structure and very light litter because of its dryness. In streams this allochthonous organic matter is an important source of energy and includes leaves, leaf fragments, floral parts, bark, branches and twigs, fruits and other plant parts. Litter also reaches aquatic habitats like streams by direct fall or lateral movement (Table 1.5).

Table 1.5. Litter input in different streams.

Stream	Stream order	Litterfall ($g^{-2}\ y^{-1}$)	Lateral movement ($g^{-2}\ y^{-1}$)	Total litter input (g-2 y^{-1})	Vegetation type
Satellite Br, N Carolina	1	35	492	629	Mixed deciduous forest
Walker Br, Tennesse	1	36	459	565	Mixed deciduous forest
Buzzards Br, Virginia	1	37	528	528	Mixed deciduous forest

Contd...

August Cr, Michigan	1	42	448	448	Mixed deciduous forest
WS10-1973, Oregon	1	45	537	1204	Coniferous forest
WS10-1974, Oregon	1	45	567	2789	Coniferous forest
Devil's Club Cr, Oregon	1	45	736	736	Coniferous forest
Rattlesnake Sp, Washington	1	47	242	242	Scrub
First Choice Cr, Quebec	1	50	417	761	Mixed deciduous forest
Breitenbach, Germany	1	51	700	700	Mixed deciduous forest
Caribou Cr 2, Alaska	1	65	37	37	Mixed deciduous forest
Caribou Cr 3, Alaska	1	65	37	37	Mixed deciduous forest
Hugh White Cr, N Carolina	2	35	506	577	Mixed deciduous forest
Deep Cr, Idaho	2	43	3	3	Shrub/ grass cover
Bear Brook, New Hampshire	2	44	594	594	Mixed deciduous forest
Beaver Cr, Quebec	2	50	217	273	Mixed deciduous forest
Monument Cr, Alaska	2	65	62	81	Mixed deciduous forest

Contd...

Creeping Swamp, N Carolina	3	35	696	696	Mixed deciduous forest
Kings Cr, Kansas	3	39	100	118	Shrub/ grass cover
White Clay Cr, Pennsylvania	3	40	313	313	Mixed deciduous forest
Mack Cr, Oregon	3	45	730	730	Coniferous forest
Keppel Cr, Australia	4	37	677	745	Mixed deciduous forest
Fort R, Massachusetts	4	42	384	384	Mixed deciduous forest
Kuparuk R, Alaska	4	70	0	500	Shrub/ sedge cover
Sycamore Cr, Arizona	5	33	17	20	Shrub cover
Kings Cr (forest) Kansas	5	39	357	726	Mixed deciduous forest
Lookout Cr, Oregon	5	45	730	730	Coniferous forest
Muskrat R, Quebec	5	50	30	41	Mixed deciduous forest
Ogeechee R, Georgia	6	32	843	4363	Mixed deciduous forest
Matamek R, Quebec	6	50	16	19	Mixed deciduous forest
McKenzie R, Oregon	7	45	218	218	Coniferous forest
Moisie R, Quebec	9	50	2	3	Mixed deciduous forest

Source: Benfield (1997)

Removal of fragmented litter is the main cause of litter disappearance from open sites (Noy-Meir, 1985; Montana *et al.*, 1988). Runoff water can also remove litter from higher points and accumulate it in lower areas during torrential rainfall events (Whitford *et al.*, 1982; Noy-Meir, 1985). The geometry of the shrubs (number, diameter and density of branches) controls the type and amount of litter retained (MacMahon and Wagner, 1985). Wind and water-flow rearrange litter in deciduous (Orndorff and Lang, 1981; Whitford *et al.*, 1982; Shure and Gottschalk, 1985) and tropical forests (Coelho Neto, 1987). Litter moves mostly down slopes and fallen logs and branches may retain litter and create a marked patchiness in the distribution of litter (Orndorff and Lang, 1981). Litter of pines tends to get accumulated asymmetrically around tree boles, and markedly so on slopes (Frankland *et al.*, 1963).

Successional trends

Accumulated litter may vary more or less regularly on successional and seasonal time scales. Olson (1963) predicted a rapid accumulation of litter during succession, followed by a steady state or quasi-equilibrium. As Net Primary Productivity (NPP) is initially higher than consumption and decomposition, there is a net accumulation of dry matter in early succession. A dynamic equilibrium reaches as NPP diminishes and decomposition rate increases (Olson, 1963; Odum, 1969). Productivity peaks associated with turnover of dominants (Whittaker, 1970), environmental fluctuations, and the existence of time lags in litter accumulation (Golley, 1965) cause less definite patterns than those predicted by most models. Perino and Risser (1972) reported that litter increases steadily along four stages (assumed to correspond to a 40 year span) of secondary succession in a tall grass prairie. Odum (1960) found that litter increased from nearly 50 g m^{-2}, in a recently cropped field, to around 350 g m^{-2} in a three year oldfield dominated by biennials, and reached to about 500 g m^{-2} when perennial grasses began to dominate. Golley and Gentry (1966) reported more dead plant material in a twelve year oldfield than in a one year oldfield community. Golley (1965) reported a six fold increase in dead plant material during 11 years of broomsedge dominance. In contrast, Wiegert and Evans (1964) found no change in the amount of litter in a herb dominated oldfield during an eleven year period. Mellinger and Mc Naughton (1975), however, did not find any evidence of litter accumulation during their study of old fields ranging in age from 4 to 36 years in central New York.

Seasonal variations

Litter production exhibits definite seasonal patterns, which vary with latitude and vegetation type (Bray and Gorham, 1964). In equatorial rain forests litter production is constant throughout the year, while in temperate evergreen forests there are peaks related to leaf production and leaf life span. In xerophytic woodlands the peak of litter production is either at the beginning or near the end of the dry period depending on the foliar strategies of the dominant populations (Madge, 1965; Hopkins, 1966).

Environmental factors not directly related to phenology also affect the seasonal patterns of litter accumulation. In two riverine forests of Belgium, woody litter increased during the winter due to the strong winds (Henry, 1987). Studies carried out by Christensen (1975) revealed that the fall of branches was determined by seasonal strong winds, and previous climatic conditions, such as water shortage during summer which predispose trees to higher litter fall. However, the seasonal peaks of litter accumulation change with successional age of the site (Perino and Risser, 1972). Odum (1960) found that seasonal differences were more conspicuous in young than in older oldfields. On the contrary, Stinner *et al.* (1984) reported more gradual litter deposition in an oldfield and in a non-tillage agro-ecosystem than in a field after conventional tillage.

Fluctuations in accumulation

Events, such as fire, wind storms, dry and hot winds, ice storms, and cold waves produce significant pulses in litter accumulation, depending on the phenologic conditions of the community and the community composition (Bruederle and Stearns, 1985). Bray and Gorham (1964) reported that species with shallow root systems shed more leaves during drought than species with deeper roots. Herbivores, such as grasshoppers that cut the leaves (Rodell, 1978) or promote leaf abscission (Choudhury, 1988; Risley and Crossley, 1988) also produce pulses of litter accumulation.

The overall dynamics of litter has been modelled by Facelli and Pickett (1991 a) to analyze the dynamics of litter at a site scale. They consider the death of plant organs produced *in situ* as the main input of litter. This input depends primarily on the complex pattern of interactions affecting the acquisition of carbon by the plant community. Of all the biomass fixed by the plants (Gross Primary Production, GPP) a portion is lost by respiration, a portion is allocated to roots, and the rest (Net Aerial Production, NAP) is allocated either to long lived or to short lived organs. The partition of biomass among different organs is determined primarily by the spectrum of life forms in the community (LFS). Before death, a proportion of biomass is consumed by herbivores. The fall of dead organs of different origin produces two different types of litter: i) persistent and ii) labile litter. Litter also accumulates by deposition of litter produced outside the system. The transformation of litter by leaching, arthropods and microorganisms, hastens (or sometimes slows) decomposition and consumption by invertebrates. Litter can also be destroyed by physical processes or removed by wind or water.

1.3. Dissolved Organic Matter

Dissolved organic matter (DOM), as stated earlier, is one of the components of detritus. It is a mixture of organic compounds ranging from simple, short-chain to complex humic substances. It is of significance is ecosystems because of its influence on acidity, microbial activity, nutrient availability and their mobility, and on the toxicity and transport of metals (Moore and Dalva, 2001). The amount of dissolved organic carbon (DOC) in the oceans is about 700 x 10^{15} g C, which is almost similar to the amount of CO_2-carbon (750 x 10^{15} g C) in the atmosphere. Soils contain about 1600 x 10^{15} g organic carbon of which a variable portion is in the dissolved form. Part of the DOC in soils (0.4 x 10^{15} g C per year) is transported across watersheds and river channels to the oceans (Amon, 2002).

Sources of DOM

Vascular plants, particularly their root exudates and microbial processes, contribute substantial amounts of dissolved C to the belowground environment (Table 1.6). In grassland ecosystems, a significant fraction (17%) of the total plant C fixation can be exuded into the soil (Biondini *et al.*, 1988). However, only a small fraction of it exists in water soluble form. Microbial control of DOC production is evidenced by several studies, but the results are somewhat contradictory. For example, studies have revealed both strong as well as weak correlations between DOC and soil respiration (Jandl and Sollins, 1997; Neff *et al.*, 2000). In the oceans and large lakes, photosynthetic algae and bacteria are the primary source of DOC. The mechanisms of phytoplankton derived production are: 1) active excretion of photosynthetic products, 2) solubilisation of senescent phytoplankton cells and sinking detritus, 3) rupture of cells due to grazing, and 4) cell lysis by pathogens. The production and subsequent release of DOM by phytoplankton is variable and depends on a variety of factors, such as growth conditions, ambient nutrient concentrations and the physiological state of the phytoplankton bloom.

Table 1.6. Nature of organic compounds identified in DOM originating from root exudates.

Major group	**Compounds identified**
Organic acids	acetic, butyric, citric, glutaric, lactic, maleic, malic, malonic, oxalic, propionic, pyruvic, succinic, tartaric, valeric
Amino acids and amides	a-alanine, β-alanine, arginine, asparagine, aspartic acid, cystine/cysteine, glutamine, glycine, histidine, lysine, methionine, phenylalanine, proline, serine/homoserine
Enzymes	amylase, invertase, phosphatase, protease, polygalacturonase
Growth factors	p-amino benzoic acid, auxins, biotin, choline, inositol, n-methyl nicotinic acid, niacin, pantothenate, pyridoxine, thiamine
Phenolic acids and coumarins	caffeic acid, cinnamic acid, coumarin, ferulic acid, salycilic acid, syringic acid, vanillic acid
Sugars	arabinose, fructose, fucose, galactose, glucose, maltose, oligosaccharides, raffinose, rhamnose, ribose, sucrose, xylose
Others	nucleotides, flavonones, fatty acids, proteins, sterols, lipids, aliphatics, aromatics, carbohydrates

Source: Koo *et al.* (2005)

Fluxes of DOM

In terrestrial ecosystems fluxes of organic compounds from throughfall and out of the litter layer amount to 1%-19% of the total litterfall C flux and 1%-5% of NPP (Neff and Asner, 2001). Fluxes of DOC generally decrease from the litter layer to deeper mineral horizons (Table 1.7). Surface soil fluxes of DOC range from 10 to 85 g C $m^{-2} y^{-1}$. but such fluxes in the sub-surface horizons decline to 2-40 g C $m^{-2}y^{-1}$ In streams, the flux of DOC ranges from 1 to 10 g Cm^{-2} y^{-1} but substantially higher fluxes can occur in drainages containing sandy or highly organic soils (Table 1.7). In general, forested ecosystems support larger fluxes of DOC than grasslands and perhaps agro-ecosystem.

Analysis of work done on DOC by Neff and Asner (2001) reveals that physical, hydrological and biological factors control the fluxes of DOC from soils. Among the physical factors, sorption dynamics and hydrology play a dominant role in regulating DOC losses from terrestrial ecosystems but the interactions between hydrology and microbial-DOC relationships are important in regulating the fluxes of DOC in the litter and sub-surface soil horizons. Sorption of DOC by soils is closely linked to soil clay content, dithionite extractable iron and oxalate extractable aluminium. Sorption isotherms are commonly used to evaluate relationships between solution concentration and soil surface association. Since DOC continuously moves in and out of solution in soils, the Initial Mass (IM) isotherm best represents DOC sorption reactions (Nodvin *et al.*, 1986) and is represented by the following linear isotherm:

$$RE = mX_i - b$$

Where, RE is the amount of DOC released into or removed from solution, m is the dimensionless regression coefficient, X_i is the initial concentration of DOC (mg g $soil^{-1}$), and b is the intercept (mg DOC released per gram soil if $X_i = 0$). Functionally, m and b can be regarded as measures of the tendency of a soil to adsorb and release DOC. IM isotherm can also be used to estimate the size of a reactive soil poll (RSP) that may be lost to leaching. This pool is determined as:

$$RSP = \frac{b}{(1-m)}$$

A synopsis of the factors controlling sorption and desorption in different soil types is given in Table 1.8. Photooxidation of DOM by UV-light has also been pointed out by several workers (Amon, 2002) but the rates of photochemical transformation of DOC under natural conditions have been rarely studied. Measurement of the extent of photodegradation of DOM in two aquatic habitats (Fig. 1.1) reveals that 15% of DOM is oxidized in Rio Niegro while more that 50% is oxidized in a lake in the Arctic Tundra.

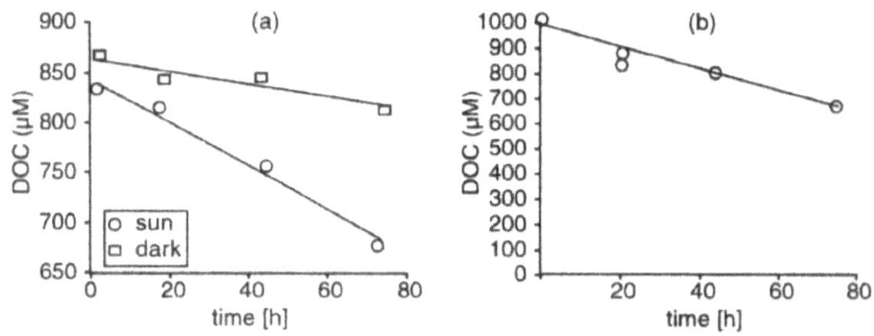

Fig.1.1. Photo-oxidation of DOM in the Rio Negro in the Amazon River system (a) and a humic lake in the Arctic Tundra (b).

Hydrological factors, though overlooked in most of the studies, also influence leaching and reactivity of DOC. Even microbial utilization and decomposition exert influence on the amount and flux of DOC in ecosystems, which, in turn are related to labile and recalcitrant fractions in the DOC pools. In general, DOC bioavailability declines as organic materials move from throughfall to deeper layers in a soil. Broad ranges of bioavailable fraction of DOC are:

Throughfall	=	18%-50%
Litter leachate	=	6%-20%
Forest soils	=	5%-16%
Agricultural soils	=	55%

In view of the variety of experimental approaches used to determine bioavailable fraction of DOC and the difficulty in evaluating relationships between DOC concentrations and decomposition rates, detailed chemical studies are needed to clarify relationships between DOC and microbial communities in soils.

Table 1.7. Fluxes of DOC across different ecosystems.

Site	Vegetation cover	DOC flux (g C m^{-2} y^{-1})	Reference
Surface soil fluxes (0-20 cm)			
Adelaide, South Australia			
Sandy loam	Eucalyptus forest and grasses	22	Steven and Cox, 1999
Clay loam	Eucalyptus forest and grasses	2-5	Steven and Cox, 1999
Coweeta Forest, NC	Deciduous forest	42	Qualls et *al*., 1991
Harvard Forest, MA	Hardwood forest	40	Currie et *al*., 1996
Harvard Forest, MA	Coniferous forest	23	Currie et *al*., 1996
Hubbard Brook, NH	Hardwood forest	21	McDowell and Likens, 1988
Medicine Bows, WY	Temperate coniferous forest	11	Yavitt and Fahey, 1986
Westlake, New Zealand	Temperate evergreen forest	84	Moore, 1989
Westlake, New Zealand	Moss/fern/ scrub forest	69	Moore and Jackson, 1989
Calhoun Forest, SC	Coniferous forest	25	Richter and Markewitz, 1996
Atlantic Plain, SC	Mixed Pine-Oak forest	13	Dosskey and Bertsch, 1997
Waldstein, Bavaria, Germany	Temperate evergreen forest	11-17	Michalzik and Matzner, 1999
Jutland, Denmark			
Heath	Heath	19	Nielsen *et al*., 1999
Oak	Hardwood forest	26	Nielsen *et al*., 1999

Contd...

Spruce	Coniferous forest	46	Nielsen *et al.*, 1999
Sub-Surface soil fluxes (20-100 cm)			
Adelaide, South Australia Sandy loam Clay loam	 Eucalyptus forest and grasses Eucalyptus forest and grasses	 2-3 3-5	 Steven and Cox, 1999 Steven and Cox, 1999
Howland Forest, ME Harvard Forest, MA	Coniferous forest Hardwood forest	3 12	Fernandez *et al.*, 1995 Currie *et al.*, 1996
Harvard Forest, MA	Coniferous forest	17	Currie *et al.*, 1996
Hubbard Brook, NH	Hardwood forest	2	McDowell and Likens, 1988
Westlake, New Zealand	Temperate evergreen forest	18	Moore, 1989
Westlake, New Zealand	Moss/fern/scrub forest	69	Moore and Jackson, 1989
Central Amazon Basin, Brazil	Tropical evergreen forest	2	McClain *et al*, 1997
Central Amazon Basin, Brazil	Tropical flooded forest	40	McClain *et al.*, 1997
Luquillo LTER, Puerto Rico	Tropical flooded forest	4-9	McDowell, 1998
Atlantic Plain, SC	Mixed Pine-Oak forest	6	Dosskey and Bertsch, 1997
Waldstein, Bavaria. Germany	Temperate evergreen forest	2	Michalzik and Matzner, 1999
Jutland, Denmark Heath Oak Spruce	 Heath Hardwood forest Coniferous forest	2 2 14	Nielsen et al., 1999 Nielsen *et al*, 1999 Nielsen *et al.*, 1999

Contd...

Stream Fluxes			
Westlake, New Zealand	Temperate evergreen forest	1	Moore, 1989
Westlake, New Zealand	Moss/fern/scrub forest	42	Moore and Jackson, 1989
Coweeta Forest, NC	Temperate deciduous forest	2	Meyer and Tate, 1983
Luquillo LTER, Puerto Rico	Tropical evergreen forest	3	McDowell, 1998

Adapted from Neff and Asner (2001)

Table 1.8. Effect of different soil attributes on sorption and desorption of DOC in different soil types.

Soil Type	**Relationship**	R^2	**Reference**
Sorption Affinity (m)			
Spodosols, Inceptisols, Alfisols, Entisols and Mollisols	m =0.64 + 0.15log Fe=0.19log Al-0.261og OC	0.75	Kaiser *et al.*, 1996
Inceptisols, Spodosols and Mollisols	m = 0.451 + 0.021log FE + 0.032"Al + 0.0641og OC	0.39	Moore *et al.*, 1992
Ultisols, Spodosols, Inceptisols and Mollisols	m = 0.15ln (% soil C) + 0.51	0.41	Neff and Asner (2001)
Desorption Parameter (b)			
Spodosols, Inceptisols, Alfisols, Entisols and Mollisols with carbonates	Log b = 0.32 + 0.72log OC	0.76	Kaiser *et al.*, 1996

Contd...

Spodosols, Inceptisols, Alfisols, Entisols and Mollisols without carbonates	Log b = -0.02 + 0.88log OC	0.82	Kaiser *et al*, 1996
Inceptisols, Spodosols and Mollisols	b = 0.145 + 0.103log OC-0.055" A1-0.045log Fe	0.72	Moore *et al*, 1992
Ultisols, Spodosols, Inceptisols and Oxisols	b = 0.05(%soil C) = 0.09	0.48	Neff and Asner (2001)

Source: Neff and Asner (2001)

Unlike grasslands or rangelands, forest ecosystems develop a thick horizon of decaying organic matter, together with highly heterogeneous accumulations of woody detritus which affect many key functions. Woody detritus reduces erosion and affects soil development; stores nutrients and water; is a major source of energy and nutrients; serves as a seedbed for plants; and is a major habitat for microbes, invertebrates and vertebrates. Despite these many functions, importance of woody detritus in forest ecosystems has, until recently, been overlooked by ecologists and forest managers (Harmon and Chen, 1991; Kirby and Drake, 1993; Samuelsson *et al*., 1994).

1.4. Woody detritus

Woody detritus includes all the forms of dead woody material above- and below-ground (Fig. 1.2). It takes many forms in forested ecosystems. For ecologists, the most useful distinctions are based on the size (length and diameter) and position (standing, downed, buried in sod) of the detritus. The most important size distinction of aboveground woody detritus is between **coarse** and **fine woody detritus fractions.** The minimum dimensions for coarse woody detritus are usually 10 cm diameter at the large end and 1.5 m in length. Smaller woody pieces than these are usually considered fine woody detritus. Coarse fractions can in turn be divided into snags (or standing dead) and logs (or dead and downed). The separation of snags from logs is usually at a 45-degree angle. Fine fractions can also be divided into suspended or downed fractions. Below-ground woody detritus has rarely been studied but distinctions into buried wood and dead coarse roots is recommended.

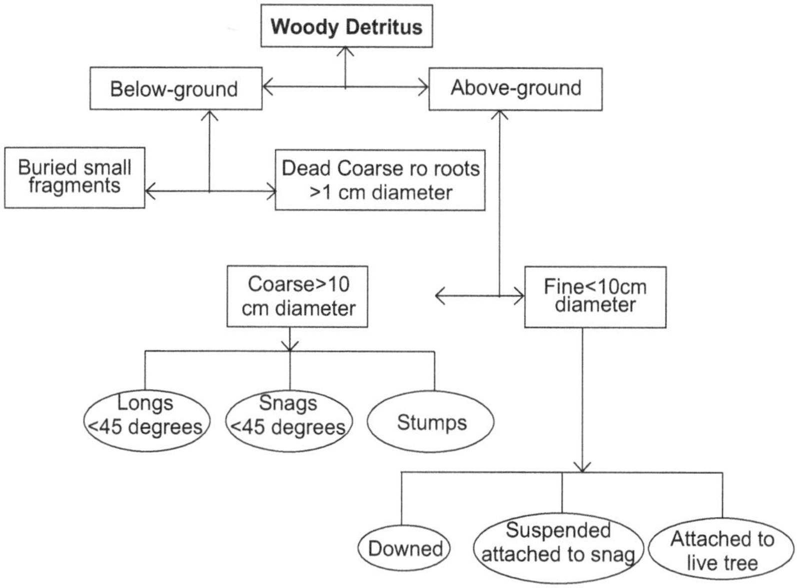

Fig.1.2. Classification of woody detritus based on position, size and degree of decomposition.

Of all the woody detritus components, coarse woody detritus (CWD) has been most widely studied, though recently several investigations have focussed on fine woody detritus also (Hely *et al.*, 2000; Klopatek, 2002; Waddell, 2002). Estimates of coarse woody detritus in forest ecosystems of different age classes are presented in Table 1.9.

1.5. Soil organic matter

Soil organic matter includes plant, animal and microbial residues in all stages of decomposition. However, some definitions of soil organic matter **(SOM)** exclude fresh plant residues from it. The division between plant residues and true SOM is apparent in dynamic models of soil C and N pools in which turnover of fresh residues are characterized

Table 1.9. Woody detritus in different forests of various age classes.

Age (range)	Location (s)	Ecosystem type(s)	CWD range (Mg C ha-1)	Reference
Boreal coniferous				
160	Finland	*Picea abies*	11.8	Finer *et al.*, 2003
25-155	Canada	*Picea mariana, Pinus banksiana*	0-6.2	Gower *et al.*, 1997
2-155	Canada	*Picea mariana*	0.0-13.0	Harden *et al.*, 1997
28-383	Siberia, Russia	*Pinus sylvestris*	4.3-26.2	Schulze *et al.*, 1999
Boreal deciduous				
53-67	Canada	*Populus tremuloides*	6.4-9.7	Gower *et al.*, 1997
5-15	Alaska	*Alnus incana*	5.2-8.2	Van Cleve *et al.*, 1971
Temperate coniferous				
Old growth	USA (Michigan)	*Tsuga canadensis*	20.1	Goodburn and Lorimer, 1998
50	USA (New Mexico)	*Pseudotsuga menziesii*	8.2	Gower *et al.*, 1992
3-1000		*Abies amabilis, Abies balsamea, Picea—Abies, Picea—Tsuga, Pinus contorta, Pinus—mixed, Pseudotsuga—Tsuga, Tsuga heterophylla, Tsuga—Picea*	1.5-24.5	Harmon *et al.*, 1986
9-316				
	USA (Oregon)	*Pinus ponderosa*	5.1-38.0	Law *et al.*, 2003
Temperate deciduous				
55	United States	*Acer saccharum—Fagus*	15.0	Borman and Likens, 1979

2-200	Chile	*Nothofagus—Drimys-Podocarpus*	9.0-206.5	Carmona *et al.*, 2002
80	USA (Indiana)	*Acer—Populus*	0.92	Curtis *et al.*, 2002
70 — old growth	USA (Wisconsin)	Northern hardwood	5.8-13.2	Goodburn and Lorimer, 1998
10-330		*Acer—Betula, Acer—Fagus, Acer—Fraxinus, Fagus—Betula, Liriodendron, Prunus pensylvanica, Quercus—mixed, Quercus—Prunus*	6.5-24.6	Harmon *et al.*, 1986
1-100	USA (Indiana)	*Quercus—Acer, Quercus—Prunus*	19.9-69.0	Idol *et al.*, 2001
8-63	USA (Wisconsin)	*Populus tremuloides*	4.3-7.8	Ruark and Bockheim, 1988
Tropical evergreen broadleaf				
5-80	Congo, Ghana, Venezuela	Dry forest, moist forest	1.25-36.0	Brown and Lugo, 1990
Undisturbed	Costa Rica	Wet forest	26.4	Clark *et al.*, 2002
10 to >150	New Zealand	*Nothofagus solandri*	11.3-87.4	Davis *et al.*, 2002
Recently disturbed—undisturbed	Mexico	Dry tropical	10.3-49.8	Harmon *et al.*, 1995
undisturbed	Brazil (Amazonia)	Rain forest	41.4	Malhi *et al.*, 1999
17 — undisturbed	Costa Rica	Secondary vegetation, wet forest	6.3-17.3	Raich, 1983
Undisturbed	USA (Hawaii)	*Metrosideros polymorpha*	0.3-2.3	Schuur *et al.*, 2001

Source: Pregitzer and Euskirchen (2004)

separately or treated as distinct pools (Wander, 2004). Although there may be a continuum of soil organic carbon compounds in respect of their stage of decomposition, physical fractionation techniques have often been used to define and delineate various discrete soil organic carbon pools (Post and Kwon, 2000) as described below.

The light fraction organic carbon **(LF-OC)** is composed of free particulate plant and animal residues undergoing decomposition and not complexed with mineral matter. In boreal and tundra ecosystems, thick surface accumulation of LF-OC occurs due to slow decomposition because of low temperature. It is a highly decomposable fraction and shows seasonal fluctuations and spatial variation with changes in litter inputs.

The organo-mineral fraction **(HF-OM)** results from the microbial transformation of soil organic carbon and its stabilization in clay- or siltsized organo-mineral complexes. Bulk of the SOC is found in this fraction.

Several workers suggest division of soil organic carbon into labile and recalcitrant fractions based on the ease with which they can be removed from the soil. Summary of the SOM pools and related organic matter fractions is given in the Table 1.10.

Table 1.10. Soil organic matter pools and related fractions.

Organic matter pools	Organic matter fractions
Labile or active SOM Material of recent origin Material of high nutrient or energy value Physically not protected and hence likely to participate in biological and chemical reactions Half-life of days to a few years	**Microbial biomass** **Labile substrates** **Litter and other residues**
Slow or intermediate SOM Physical status, physical protection, or location is used to separate this fraction from other two fractions Half-life of a few years to decades	**Partially decomposed residues and decay products** **Some humic materials**

Recalcitrant, passive, stable and inert SOM Recalcitrance because of biochemical characteristics and/or mineral association	**Refractory compounds of known origin** **Some humic substances**

Balance between the rate of soil organic carbon inputs and rate of mineralization determines the amount of organic carbon stored in soil. Accumulation of organic carbon in Holocene age soils has been compiled by Schlesinger (1990). He documented long-term rates of carbon storage from 0.2 g $Cm^{-2}y^{-1}$ in some polar deserts to greater than 10 g $Cm^{-2}y^{-1}$ in some forest ecosystems, with an average rate of 2.4 g $Cm^{-2}y^{-1}$ over all ecosystems. Cao and Woodward (1998) used **CEVSA** (Carbon Exchange between Vegetation, Soil and the Atmosphere) model to estimate the amount of soil organic carbon in different terrestrial ecosystems (Table 1.11). This study reveals that a total of 1358 Gt C is present in the world's soils. Boreal woodland and forests hold 27%, and tropical forests and savannah hold 23% of the total.

Table 1.11. Estimates of soil carbon content in different terrestrial ecosystems of the world using CEVSA model.

Vegetation type	**Area** (10^6 km^{-2})	**Soil C content** (kg m^{-2})
Polar desert and alpine tundra	4.9	0.5
Wet and moist tundra	4.7	0.7
Boreal woodland	6.3	1.3
Boreal forest	12.1	1.2
Temperate coniferous forest	2.3	0.9
Desert	11.1	0.0
Arid shrubland	14.4	0.3
Short grassland	4.5	0.7
Tall grassland	3.5	0.8
Temperate savannah	6.5	0.6
Temperate mixed forest	4.9	0.9
Temperate deciduous forest	3.4	1.0

Temperate broadleaf evergreen forest	3.1	0.9
Mediterranean shrubland	1.3	0.4
Tropical savannah	14.0	0.5
Xeromorphic forest	6.8	0.4
Tropical deciduous forest	4.6	0.6
Tropical evergreen forest	17.4	0.8

Source: Cao and Woodward (1998)

•••

CHAPTER 2

EFFECT OF DETRITUS ON ENVIRONMENT

1.2. The physical environment

Presence of litter alters the micro-environmental conditions of top soil by intercepting incident light and rain and also by affecting the transfer of heat, water and gases between the soil and the atmosphere. Such alterations have a bearing on the plant community structure directly through effect on the germination and establishment of plants and indirectly through changes in the resource availability and other biotic components.

Light

Interception of light by litter is of importance in open systems, such as old fields and grasslands, as well as in systems with seasonally closed canopies, such as deciduous forests (Ellenberg, 1988). Total radiation below a dense mat of grass litter may be only 1% to 5% of the radiations present above the litter (Knapp and Seastedt, 1986). Shading by litter follows the light extinction law of Beer-Lambert, with an exponential reduction of the light interception coefficients. Effect of litter on light environment is also a function of its weight/surface ratio (Sydes and Grime, 1981b). Variable effects of litter of broad-leaved trees and conifers on the ground vegetation is partly due to their different shading properties (Ellenberg, 1988; Facelli and Pickett, 1991b) and partly due to changes in the spectral composition of light especially when litter is moist (Vazquez—Yanes *et al.*, 1990). In fact, canopy foliage reduces the red to far-red ratio (r:fr) of light reaching the ground surface by preferential absorption of light in the wavelengths of 660 nm (red light) relative to longer wavelengths (far-red light). For the same reason, dead leaves on the forest floor decrease the r:fr ratio of light reaching seeds buried beneath them (Pearson *et al.,* 2003). However, litter of different species show significant differences in the r:fr ratio of transmitted light (Table 2.1).

Table 2.1. Effect of fresh and dried leaf litter of different species on red: far-red ratio of transmitted light in comparison to bright sunlight (r:fr ratio 1.36).

Species	Fresh litter	Dried litter
Miconia argentea	0.55	0.47
Ochroma pyramidale	0.71	0.49
Luehea seemannii	0.35	0.52
Solanum hayesii	0.81	0.79
Anacardium excelsum	0.30	0.18
Cecropia spp.	0.41	0.83

Source: Pearson *et al.* (2003)

Such changes in the light environment have important ecological effect on seed germination (Fenner, 1985), seedling development (Ballare *et al.*, 1988) and tillering (Deregibus *et al.*, 1985). Particularly affected are those plants whose seeds respond positively to light (Grime, 1979; Sydes and Grime, 1981b; Vazquez-Yanes *et al.*, 1990). Light deprivation of seedlings by litter causes their mortality due to extra energy cost of seedlings to penetrate the litter mat. Etiolation of seedlings growing under litter is also a cause of seedling mortality. Inhibition of growth of the tree seedlings (Tao *et al.*, 1987) and increased damping off of Douglas fir seedlings (Herman and Chilcote, 1965) due to shading by litter is already on record.

Even if seedlings survive after expending considerable energy to reach the surface, their tillers show morphological and physiological characters that decrease their capacity to fix carbon (Knapp and Seastedt, 1986) thus affecting primary productivity and other allied ecosystem functions.

Soil temperature

Litter modifies soil temperature by intercepting solar radiation, and by insulating the soil (Evans and Young, 1970) due to which it delays freezing of soils during cold seasons. Weaver and Rowland (1952) found that soil temperature in a 'mulched' grassland was 8°C lower than that in an unmulched one, and that the temperature of air was higher than that of the soil surface under the litter. Burning, as well as, hand removal of litter also increases soil temperature (Hulbert, 1969). Unburied crop residues produce similar effects on the soil temperature (Holland and Coleman, 1987). Such changes in soil temperature produced by litter directly affect plant growth and also enhance the rates of mineralization, hence nutrient availability (Chapin *et al.*, 1979; Knapp and Seastedt, 1986). Plants, in grasslands with dense litter accumulation, grow more slowly and flower more sparsely due to lower soil temperatures during spring (Weaver and Rowland, 1952; Rice and Parenti, 1978). For example, accumulation of dead materials in the crown of the tundra sedge *Eriophorum vaginatum* provides insulation that allows for an extended growing season and faster nutrient cycling, thus improving productivity (Chapin *et al.*, 1979). Protection against frost heaving by litter is discernable in early successional communities where tree establishment is negligible before litter accumulation (Small *et al.*, 1971; Pickett *et al.*, 1987; McCarthy and Facelli, 1990). Differences between plant communities of temperate forests found in pits and mounds are partly explained by differences in soil temperature resulting from differential litter accumulation in the two microhabitats (Beatty and Sholes, 1988). Reduction of soil thermal amplitude produced by litter has also been shown to impair germination of seeds requiring alternating temperature regimes (Thompson *et al.*, 1977; Grime, 1979; Fenner, 1985) and this mechanism has been invoked for the local extinction of many weedy species from old fields and ungrazed grasslands (Facelli *et al.*, 1987).

Water dynamics

Litter affected exchange of water between soil and atmosphere has been observed in grasslands, deserts and floodplains (West, 1979; Fowler, 1986; Eckstein and Donath, 2005). The accumulation of litter decreases run off, reduces the impact of rain drops on the soil and prevents disaggregation (Dyksterhuis and Schmutz, 1947). It also induces changes in soil pH that affect physical structure of the soil, and therefore, water percolation properties as well (Walsh and Voight, 1977). In grasslands and crop fields, litter on the soil surface increases water availability due to reduced evaporation (Holland and Coleman, 1987) by directly increasing the resistance to water vapour diffusion from the soil surface and, indirectly by reducing the soil temperature (Facelli and Pickett, 1991a).

Under certain circumstances water retention by litter may reduce its availability to plants (Walsh and Voight, 1977; Knapp and Seastedt, 1986). As much as one third of daily rain may be retained by litter followed by its direct evaporation without becoming available to the plants (Weaver and Rowland, 1952). Tao *et al.* (1987) also reported higher moisture content in the Korean pine forest plots without litter than those with intact litter cover. However, net effect of litter on the water balance depends on its water retention potential and also on the rainfall pattern. For instance, when most of the rain falls in small rainfall events, proportion of water retained by the litter is larger than that retained if the same amount falls in a few large events (Walsh and Voight, 1977; Sala and Lauenroth, 1983). Such changes in water availability mediated by litter have obvious effects on germination, seedling establishment, growth and other related attributes of plants.

2.2. Chemical environment

Different organic and inorganic substances are accumulated by plant organs during the entire course of their growth and senescence. After death of these organs or sometimes during senescence, these compounds are either released by leaching or their decomposition products are released into the soil. The chemical properties of such leachates (e.g. toxicity, mineral contents, solubility etc.) depend on the nature of the substances accumulated in the organ before its fall and on the biochemical transformations brought about by decomposers. Thus, depending on the properties, these chemical compounds alter community structure and also affect the availability of nutrients (see chapter 8) for uptake by plants.

Production of phytotoxins

There is abundant literature dealing with the production of phytotoxic compounds by litter upon its leaching or decomposition (Garnett *et al.*, 2004). However, it needs to be emphasized that microorganims have the potential to convert nontoxic compounds to toxic ones (e.g. amygdalin in peach residues) or synthesize inhibitors while decomposing residues of plants (e.g. patulin by *Penicillium urticae*). Though exact mechanisms of action are poorly understood, negative effects of litter leachates on germination and growth are extensive (Welbank, 1963; Abdul-Wahab and Rice, 1967; Rice, 1979; Carter and Grace, 1986; van der Valk, 1986; Weir *et al.*, 2004). Most of such evidence is based on laboratory or greenhouse experiments wherein addition of litter (or litter leachates) reduces the germination and/or the growth of seedlings in different species (Rice, 1979). However, effect of litter leachates on the community structure has seldom been tested in the field and hence the role of chemical compounds that leach out of the litter in organization of natural communities remains unassessed.

Notwithstanding their release from litter, phytotoxic effects of litter leachates have been questioned (De Jong and Klinkhamer, 1985; Stowe, 1979) because some field studies (Collins and Quinn, 1982; De Jong and Klinkhamer, 1985; Carter and Grace, 1986) present inconclusive results while others (Rice and Parenti, 1978; West, 1979) even show negative results. Besides, in many of the experiments fresh litter or even the live clipped organs have been used, which according to Facelli and Pickett (1991 a) is unrealistic. For example, Bokhari (1978) reported that extracts of fresh leaf litter reduce the germination of grasses more than does old leaf litter (Sydes and Grime, 1981b). Sydes and Grime (1981b) found that the initial negative effects produced by the litter leachates on seedling size disappeared long before the herbs reached the reproductive stage. In addition, the litter types that produced more phytotoxicity were those that decomposed faster, and therefore, the effect presumably should disappear rather rapidly (Sydes and Grime, 1981 b). Similarly Szczeponska (1977) reported that though litter addition initially reduced the growth rates of both *Phragmites communis* and *Typha latifolia*, but after three weeks such litter induced changes in growth rates did not persist. However, differences in biomass of the two species did persist (Carter and Grace, 1986).

2.3. Biological environment

Litter influences, directly as well as indirectly, many attributes of the ecosystems, such as community structure, trophic dynamics and biodiversity.

Trophic dynamics

Detritus, a central component of food webs in terrestrial, freshwater, and marine ecosystems (Moore *et al.*, 2004), provides energy base for detrital food web loops (Hall *et al.*, 2000). It links decomposers to producers directly via mycorrhizal mutualisms (Reid and Woods, 1969; Fogel, 1980; Fogel and Hunt, 1983; Stjohn *et al.*, 1983) and indirectly via nutrient mineralization associated with decomposition in soils or in the water column (Wardle, 1999). Coupling of food webs in above- and below-ground compartments of terrestrial ecosystems (Wardle, 2002) as well as between aquatic and terrestrial ecosystems (Vannote *et al.*, 1980; Polis *et al.*, 1997) is also brought about by detritus.

Although detritus serves as both food and habitat for members of detrital as well as autotroph based compartments of food webs but more importantly, coarse detritus, such as large woody debris, provides key habitat for detritivores (Wallace and Benke, 1984; Benke and Wallace, 1990) as well as predators (Harmon *et al.*, 1986; Fausch and Northcote, 1992; Everett and Ruiz, 1993; Steel *et al.*, 1999). For example, leaves and associated microbial and fungal decomposers are the primary source of carbon for invertebrate and vertebrate detritivores living either on the soil surface and/or in deeper soil horizons (Anderson, 1975; Mitchell and Parkinson, 1976; Swift *et al.*, 1979; Ponge, 1991; David *et al.*, 2001; Hattenschwiler and Bretscher, 2001). Leaf litter food webs exemplify the inherent donor-controlled nature of detrital consumer-resource systems (Polis and Strong, 1996). Like coarser forms of detritus, leaf litter can also determine the relative abundance of species that are not dependent on litter as an energy source, like ground spiders which have a strong affinity for leaf litter on forest floors (Uetz, 1979; Wise, 1995; Vargas, 2000; Wagner *et al.*, 2003). These predators may depend on the structural complexity of litter as cover from other predators (Uetz, 1979; Vargas, 2000). Depth and complexity of litter also alter the thermal environment of surface and deeper soil horizon, which in turn influences the distribution and abundance of spiders (Rypstra *et al.*, 1999) and other invertebrate species. Finally, decomposition of litter by soil flora and fauna directly influences local pools of nutrients available for plant production (Ingham *et al.*, 1985, Setala and Huhta, 1991), and the activities of predators can, in some cases, depress decomposer populations leading to inhibition of decomposition (Wise *et al.*, 1999; Lawrence and Wise, 2000; Rosemond *et al.*, 2001). Thus, detritus not only affects the trophic structure and dynamics of communities but also influences the species diversity and length of food chains. However, the role of living plant organic matter across trophic levels has been overemphasized when the detritus, a common component of most ecosystems, is largely overlooked in the study of food webs and trophic dynamics. Deconstruction of 40 community food webs compiled by Briand (1983) into 138 pathways or energy channels that originate with a basal resource (e.g. detritus, primary producers, or consumers) and end with a top predator (Moore and Hunt, 1988) revealed that majority (72.5%) of the community webs contained both detritus and primary producer energy channels. Of the 138 energy channels, 63% originated with a primary producer, 20% with detritus, and 17% with consumers; many of which could be traced back to detritus if the descriptions were complete. Alsothe amount of energy that flows through the detrital pathway can equal or exceed that of the grazing branch (Hairston and Hairston, 1993; Wetzel, 1995; Cole *et al.*, 2000; Cole and Caraco, 2001; Heymans *et al.*, 2002; Mulholland *et al.*, 2002) and some food webs, such as those in caves, small streams in forested watersheds, and below-ground are based almost entirely on detritus (Hunt *et al.*, 1987; Wallace *et al.*, 1997; Jesser, 1998). In autotroph based food webs also, the detritus pathway can have strong influences on the structure and dynamics of the grazer pathway by providing energy that can sustain higher densities of consumers than would otherwise be maintained if these consumers fed exclusively on energy derived from the grazer pathway (Polis *et al.*, 1997; Moore *et al.*, 2003). For example, detrital inputs often translate into increased biomass of spiders and lizards on oceanic islands; this high number of predators can suppress herbivores with subsequent effects on plant abundance and species composition (Polis and Hurd, 1996). In this case, both the trophic cascade from predators to plants and average lengths of the food chains are driven by detrital inputs from outside the island ecosystem. Likewise, in the reservoirs of midwestern USA, detritivorous fish (gizzard shad) attain high biomass via consumption of allochthonous and autochthonous detritus and excrete nutrients that stimulate phytoplankton biomass (Schaus and Vanni, 2000; Vanni, 2002).

In short, linkages between detritus and grazer pathways are many and varied (Moore *et al.*. 1988; Polis and Hurd, 1996; Polis *et al.*, 1997; Azam, 1998), but often occur within two trophic levels from either source (Table 2.2). Studies have revealed that the strength of these connections varies among and within ecosystems. For example, in oceanic islands top predators, such as spiders and lizards link the two pathways, while in many reservoirs the pathways are strongly linked by a single species of fish. Despite the empirical evidence presented, grazer and detritus pathways are often treated separately, or when depicted together they are not connected in a significant way.

Table 2.2. Percentages of trophic species that occupy >2nd trophic level (TL) and the top trophic level and obtain energy from detritus.

Food web	% TL >2	%Top TL	Reference
US Marine shelf	97.73	100	Opitz, 1996
Caribbean reef	97.01	100	Link, 2002
Chesapeake bay	100	100	Baird & Ulanowicz, 1989
St Marks estuary	100	100	Christian & Luczkovich, 1999
Little Rock lake	96.97	100	Martinez, 1991
Stony stream	100	100	Townsend *et al.*, 1998
Canton Creek	100	100	Townsend *et al.*, 1998
St Martin Island	96	100	Goldwasser & Roughgarden, 1993
Coachella Valley desert	100	100	Pollis, 1991
Shortgrass steppe soil	100	100	Hunt *et al.*, 1987
Toolik Lake tundra soil	100	100	Doles, 2000
Lovinkhoeve farm soil	100	100	De Rutter *et al* 1995
Horseshoe Bend farm soil	100	100	Hendrix *et al.*, 1986
Kjettslinge farm soil	100	100	Andren *et al.*, 1990

Source: Moore *et al.* (2004)

Community structure

The changes induced by litter in the physical and chemical environment affect different demographic processes (Xiong and Nilsson, 1999; Xiong *et al.*, 2003; Rotundo and Aguiar, 2005) and also the outcome of interactions between populations. Such alterations eventually affect the community structure. Major effects of litter on various demographic processes are summarized in the Fig. 2.1.

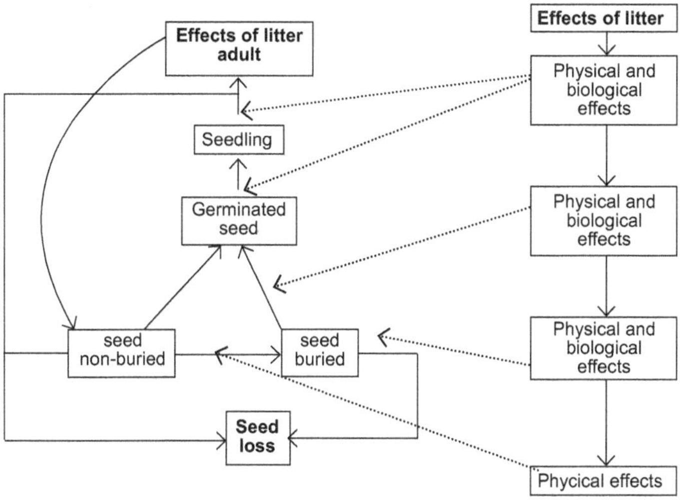

Fig. 2.1. Summarized representation of the effects of litter on demographic processes.

Germination and establishment, two key factors in plant community organization (Grubb, 1977), are particularly sensitive to the presence of litter (Smith, 1995; Kozlowski, 2002; Makana and Thomas, 2005). It enhances the establishment of some species by improving water conditions (West, 1979; Fowler, 1986) or by reducing competition while establishment of many species is negatively affected because of litter induced shading, mechanical impedance, reduced soil thermal amplitude or biochemical effects. Several observational as well as experimental studies on temperate and tropical trees have indicated that exposed mineral soil improves seed germination and seedling establishment (Karlsson and Orlander, 2000; Harrington and Bluhm, 2001). Compared to litter, exposed mineral soil is regarded as a better seedbed because it provides good hydraulic contact between the seeds and soil particles.

On the contrary, litter impedes the penetration of roots, thereby preventing contact with mineral soil (Kozlowski, 2002). However, some workers (Molofsky and Augspurger, 1992) argue that the exposed mineral soil may be more important for small-seeded species rather than for species with large seeds. For example, establishment of Prunella vulgaris is enhanced by herb litter in old fields, but in forest habitats its establishment is reduced by leaf litter (Marks, 1983; Winn, 1985). Watt (1974) observed that most annuals were excluded from chalk grassland by a dense litter layer that had accumulated due to absence of grazing animals. Collins and Good (1987) reported that litter was an important component of the regeneration niche of various species in the New Jersey pine barren systems. Similarly, the distribution of tree species in southern Pennsylvania is known to be affected by the ability of their seedlings to tolerate the accumulation of litter (Keever, 1973). In the highland grasslands of Victoria, Australia, shrubs establish only in patches where grass litter is removed by various disturbance agents (Williams and Ashton, 1987). Likewise, Hamrick and Lee (1987) suggested that the absence of musk thistle (*Carduus nutans*) in well managed pastures could be attributed to the accumulation of grass litter. In contrast, the establishment of some weeds of arid rangelands (*e.g. Bromus tectorum, Taeniatherum aspersum, Salsola kali*) was enhanced by litter accumulation because of improved water availability (Evans and Young, 1970; Evans, 1972).

Meta-analysis of 35 independently published studies carried out by Xiong and Nilsson (1999) reveals that the litter has an overall Negative effect on the germination and establishment of plants (Table 2.3). However, magnitude of the effect varies with latitude, habitat, target species, quantity and type of litter.

Table 2.3. Overall effect of litter on seed germination and establishment of plants.

Demographic process	Effect size	Confidence Interval	Probability	Number of studies
Germination	-0.60	-0.71~-0.49	0.01	160
Establishment	-0.42	-0.57~-0.28	0.01	52

Source: Xiong and Nilsson (1999)

The overall negative effect of litter on the two important demographic processes of germination and establishment implies that vegetation development is dependent on the litter reducing factors, such as decomposition and disturbances (burning, flooding, animal trampling etc). Inhibition of seed germination more than establishment is suggestive of importance of disturbance at the beginning of the growing season when it coincides with germination than at later stage of plant establishment. Significant variation in the effect size of litter on germination and establishment with latitude is consistent with the fact that natural litter accumulation increases with latitude (Vogt *et al.*, 1986). The study further reveals significant correlation between litter quantity (measured as litter depth) and seed germination and between establishmeot and litter mass (g m^{-2}) as well as its depth. Litter type also significantly influences germination and establishment; with forb litter producing more negative effect on germination and evergreen litter on establishment. However, grass litter does not exhibit any influence on seed germination but has marginal effect on establishment. Such variable effects of litter type are a result of different light, temperature and moisture conditions and also due to chemical differences in nutrient availability and allelopathy. Germination and establishment responses to litter also vary when comparisons are made across ecosystems. Litter effect on germination is significant in old fields and forests but not in deserts and grasslands while as effect on establishment is negative in all ecosystems except deserts due to conservation of moisture but this positive moisture-conservation effect is counterbalanced by a negative chemical impact on seed germination. Trees, grasses and forbs also differ with respect to effect of litter on their germination and establishment as these two demographic processes are more negatively affected by litter in trees than in forbs and grasses. Litter accumulation is also reported to affect productivity, diversity and species composition (Knapp and Seastedt, 1986). Sydes and Grime (1981a) found a consistent correlation between shoot biomass and persistent litter in forest floor communities, and suggested that litter is the main factor controlling the structure of the communities; the tenet stressed by Ellenberg (1988) and Hermy (1987). Persson *et al.* (1987) attributed changes in the ground forest vegetation layer over 50 years to changes in litter composition from predominantly *Quercus* and *Corylus* litter to litter of *Ulmus* and *Fergus*. In the Californian annual grassland, litter removal decreased productivity (Heady, 1956) probably because of deterioration in water conditions. Meta-analytical approach employed by Xiong and Nilsson (1999) also indicated significant influence of litter on above-ground biomass (Table 2.4). However, effect the switches from negative to positive over a longer time period possibly due to recovery of vegetation from burial under litter and release of nutrients following litter decomposition.

Table 2.4. Overall effect of litter on above-ground biomass and species richness

Vegetation variable	Effect size	Confidence Interval	Probability	Number of studies
Above-ground biomass	-0.18	-0.34~-0.06	0.05	34
Species richness	-0.63	-0.84~-0.43	0.01	25

Source: Xiong and Nilsson (1999)

Relationship between litter and above-ground biomass also shows a definite latitudinal pattern of more negative effect of litter in high latitudes which is consistent with global pattern of litter decomposition (Berg *et al.*, 1993). Thus, more the litter accumulation, greater is its negative impact on the above-ground biomass.

Biodiversity

As early as 1967, population biologists pointed out that the number of species in a community simply depends on the rate at which species disappear from the vegetation and the rate at which new species establish (MacArthur and Wilson, 1967). Studies have shown that accumulation of plant litter generally reduces species richness in plant communities (Xiong and Nilsson, 1999), such as grasslands (Penfound, 1964; Watt, 1974; Facelli *et al.*, 1988), old fields (Facelli *et al.*, 1987; Carson and Peterson, 1990), lacustrine wetlands (van der Valk, 1986), salt marshes (Haslam, 1971) and in pit and mound complexes in temperate forests (Beatty and Sholes, 1988). Factors resulting in removal of litter may, therefore, be of importance is contributing to species richness in productive communities. Experiments have actually shown that litter removal increases species diversity and the number of flowering species in grasslands (Weaver and Rowland, 1952; Penfound, 1964). Haslam

(1971) reported that litter accumulation in *Phragmites* communities maintained dominance of the reed and prevented invasion by other species. In a five year successional study in the Argentine pampa, reduction of the species diversity and the local extinction of ruderal species were seemingly related to the ability of *Loliwn multiflorum* to accumulate a dense layer of litter (Evans and Young, 1970; Facelli et *al*., 1988). Contrarily, evidence for reduced dominance and increased diversity in the ground herb community of forests due to litter accumulation is also available (Sydes and Grime, 1981b) and such a response has been attributed to the release of dicots from competition by dominant grasses. While separating the effects of increased living biomass production from those of increased litter production (resulting from increased nutrient supply) on the species richness, Foster and Gross (1998) found that nitrogen fertilization strongly increases both living plant biomass and litter biomass leading to reduced species richness by preventing seedling establishment of the subordinate forbs. In comparison, litter removal results in an increase in species richness and forb establishment by the same amount in fertilized and unfertilized treatments, indicating thereby that factors other than nutrient release from litter decomposition are operative. Many works pertaining to the study of nutrient supply on the diversity of plant communities have focussed on how nutrient inputs affect competition between already established plant individuals, but few workers have attempted to analyze experimentally the effects of nutrient supply on the establishment of plant species which is of paramount importance, especially for short lived species which strongly depend on continuous reestablishment (Berendse, 1999). In general, it has been accepted that species diversity in many plant communities declines as nutrient supply increases (Heij and Schneider, 1991). The traditional point of view is that increase in nutrients permits few fast growing, tall plant species to profit from the additional nutrients which expand rapidly and out shadow the variety of species that have lower potential growth rate, or that, for other reasons are unable to profit from extra available nutrients (Grime, 1979; Huston and DeAngelis, 1994). Such a relationship between litter accumulation and species richness has been shown to be determined by direct and indirect interactions between a variety of factors, such as elevation, stage of vegetation development, vegetation canopy and availability of propagules, etc. (Xiong *et al*., 2003).

The above discussion allows to conclude that litter is an important factor affecting community organization and dynamics far beyond its commonly recognized role as a transitory bank of nutrients. The manifold effects of litter on different responses of various variables and the different responses of various populations to its presence preclude a general prediction of the effect of accumulated litter on community structure (Bakker, 1985; Facelli, 1988; Facelli and Pickett, 1991a).

•••

CHAPTER 3

DECOMPOSITION: PROCESSES AND PATTERNS

3.1. Processes

Decomposition is defined as the physical and chemical breakdown of detritus. This detritus breakdown releases carbon to the atmosphere and nutrients into soils in the forms that can be used by plants and microbes and also results in a pool of complex organic compounds (humus) that are resistant to further microbial breakdown (Chapin *et al.*, 2002). Plant litter decomposition signifies the mechanical disintegration of dead plant structures from the stage where they are still attached to the living plant to the humus stage where the gross cell structure is no longer recognizable (Satchell, 1974). It causes a decrease in detrital mass as materials are converted from dead organic matter into inorganic nutrients and CO_2 (Guzman, 1997).

Above statements bring out that decomposition is a complex and often prolonged process and its rate is controlled by the nature of the substrate and characteristics of the environment (Satchell, 1974; Singh and Gupta, 1977; Smith and Bradford, 2003). Detritus decomposition also encompasses interactions among many types of organisms and environmental variables (Kurihara and Kikkawa, 1986; Gunnarsson *et al.*, 1988), each affecting quality of the litter as a substrate for the subsequent decomposers (Wiegert *et al.*, 1970; Seastedt and Crosley, 1983; Richards, 1987; Choudhury, 1988; Gunnarsson *et al.*, 1988; Homer *et al.*, 1988). For example, in seagrass litter (*Zostera marina*), leaching and volatilisation (increased by fragmentation) change the substrate quality (Kenworthy *et al.*, 1987), making litter available for invertebrates and decomposers (Harrison and Mann, 1975; Wahebh and Mahasheh, 1985). Likewise, size of the particles left by litter consumers determines which organisms are likely to attack the material subsequently (Furniss and Ferrar, 1982; Richards, 1987; Hagvar, 1988). Also in the absence of earthworms or other invertebrates that fragment litter, microbial decomposition is very slow and litter persists longer (Lee, 1985).In fact studies conducted in sub-humid and arid systems where earthworms are scarce or absent suggest that physical fragmentation is needed to make litter available to arthropods and microbes (Noy-Meir, 1985; Montana *et al.*, 1988).

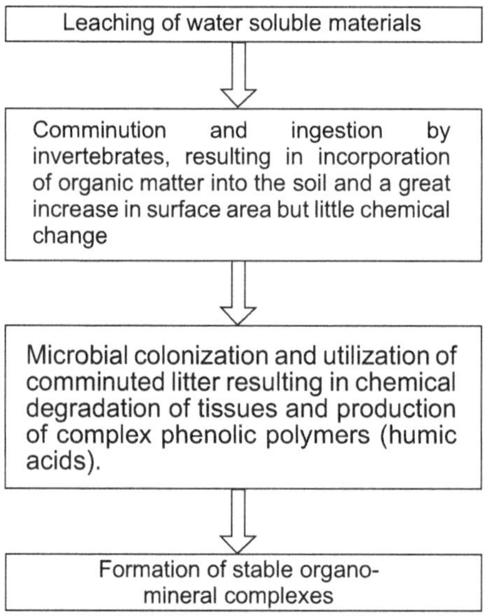

Fig. 3.1. Stages of detritus decomposition.

The process of litter decomposition is usually divided into the following three sub-processes having different controls and consequences (Mason, 1977; Chapin *et al.*, 2002) (Fig. 3.1):

- Leaching
- Fragmentation / Comminution
- Catabolism / Chemical alteration

Leaching

It is the physical process by which mineral ions and small water soluble organic compounds dissolve in water and move through the soil resulting in weight loss and change in chemical composition of the litter (Swift *et al.*, 1979). Several workers have demonstrated the importance of leaching in litter decomposition (Nykvist, 1961; Anderson, 1973a; Day, 1983; Gupta and Arun Lekha, 1989). Gupta and Singh (1981a) reported that for five different plant materials, the mass loss due to leaching ranged from 26.1% to 75.5%. Similar results were obtained by Gupta and Arun Lekha (1989). Rajvanshi and Gupta (1980) reported 27.9% mass loss from litter bags due to leaching in case of *Dalbergia sissoo* leaves. Similarly, Arun Lekha (1987) found 14% to 27% mass loss in litter of different species due to leaching within 48 hours of the placement of litter containing bags in water. In aquatic ecosystems as well, leaching is considered to be the characteristic mechanism initiating leaf breakdown that leads to substantial mass loss (up to 30%) within 24 hours after immersion of leaves (Peterson and Cummins, 1974; Benfield, 1997). Factors like water temperature, turbulence and nature of the detritus determine the extent of mass loss (Webster and Benfield, 1986). When decaying leaves are ground for analysis (Suberkropp *et al.*, 1976; Paul *et al.*, 1983) or fragmented by invertebrate feeding (Meyer and O'Hop, 1983), additional soluble organics are released resulting in further mass loss.

On a global scale, considerable variation in leaching behaviour of detritus is determined by canopy architecture, timing of leaf fall, prevailing weather conditions, stream channel and bank valley morphology (Gessner, 1999). Leaves of some species die naturally on the trees and leaching in such leaves occurs instantly upon their immersion in water. Such leaves may also loose solutes during rain even before abscission. In other situations leaves may be initially trapped in the canopy (Covich, 1988; Campbell and Fuchshuber, 1994) or on the ground (Mayack, *et al.*, 1989) where they may undergo partial breakdown and loose their solutes before they enter aquatic environment.

Leaching, in fact, is the rate determining step for mass loss of litter when it first falls to the ground (Chapin *et al.*, 2002). During senescence, many of the compounds in a leaf are broken down and transported to other plant parts. This resorption process is still actively occurring when the leaf is shed, so senesced leaf contains relatively high concentrations of water soluble products that are readily leached. Thus, leaching begins when tissues are still alive and is most important during tissue senescence and when litter first falls to the ground. Leaching losses from fresh litter are greatest in environments with high rainfall and are negligible in dry environments (Cadish and Giller, 1997). Compounds leached from leaves include sugars, amino acids and other compounds that are labile (readily broken down) or are absorbed intact by soil microbes. Litter leachates frequently support a pulse of microbial growth and respiration during periods of high litter fall. It should be noted that leaching transfers the soluble resource material to a different site usually lower down the vertical profile of the ecosystem where it may subsequently undergo further decomposition (Swift *et al.*, 1979).

Transport of detrital particles, particularly significant in aquatic ecosystems, is variable, depending on the size and density of particles and on physical characteristics of the stream including depth, stream power and the abundance of retentive structures (Webster *et al.*, 1999). Large logs that fall into small streams rarely move at all, whereas they may move a few meters in large streams before accumulating the debris; large rivers may transport logs all the way to the ocean. The distances travelled by small sticks similarly vary with stream size. Leaves generally move only a few meters before hitting obstructions (Webster *et al.*, 1994) and often stay at a single spot until they are broken down, although they may be transported downstream to short distances during high flow. The fine particulate organic matter (FPOM) obtained from biotic fragmentation, foliar tissue degradation by microbial pectinases (Suberkropp and Klug, 1980; Ward *et al.*, 1994; Webster *et al.*, 1995), conidia of aquatic hyphomycetes (Barlocher, 1982; Suberkropp, 1992b; Gessner and Chauvet, 1994), detached bacteria (Findlay and Arsuffi, 1989; Weyers and Suberkropp, 1996; Baldy and Gessner, 1997) and hyphal fragments (Ward *et al.*, 1994) are transported in a series of saltations. Once particles are in water column they may be transported to several hundred meters depending on size of the stream, characteristics of substratum and the size and density of the particles. The residence time of these particles is highly variable after retention on the bottom. However, resuspension of a particle generally requires an increase in stream flow (Webster *et al.*, 1999).

Fragmentation/Comminution

It is a physical process brought about by feeding activity (both ingestion and digestion) of decomposer animals leading to reduction of particle size of organic resources (Swift *et al.*, 1979). Fresh detritus is initially covered by a protective layer of cuticle or bark and by skin or exoskeleton in animals to protect tissues from microbial attack (Chapin *et al.*, 2002). Fragmentation increases the proportion of the litter mass that is accessible to microbial attack by creating fresh surfaces for microbial colonization and greatly enhances microbial decomposition by piercing these protective barriers and by increasing the ratio of litter surface area to mass. Although, animals are the main agents of litter fragmentation, freeze-thaw and wetting-drying cycles also disrupt the cellular structure of litter. In aquatic ecosystems, particularly streams, physical fragmentation of decomposing leaf litter occurs as a result of abrasion and shear stress exerted by the flowing water (Gessner, 1999) but probably it is less important in lakes and tidal marshes (Odum *et al.*, 1972). A second mechanism of fragmentation (comminution) is promoted by invertebrates and is termed as biotic fragmentation (WebsterandBenfield,1986).Itisgenerallyascribedtothecomminution of leaves through the feeding and digestive activities of shredders, resulting in the release of fine particulate organic matter (FPOM) (Cummins, 1974; Cuffney *et al.*, 1990; Allan, 1995). These shredders preferentially colonize and feed on microbially conditioned leaves and may contribute significantly to leaf breakdown in streams (Peterson and Cummins, 1974; Iversen *et al.*, 1982; Wallace *et al.*, 1982). Conditioning means "preparation" of leaf material for the invertebrate consumer. This process of conditioning enhances litter fragmentation through its influence on leaf litter palatability to detritivorous macro-invertebrates(shredders) due to the accumulation of microbial biomass of high nutritive value (Cummins, 1974). The concept thus suggests alinear cause and effect relationship in which microbial colonization facilitates shredder feeding and therefore leaf mass loss (Gessner, 1999). As a matter of fact leaves are colonized by a variety of aquatic microbes within a few days of deposition in freshwater. Fungi, principally hyphomycetes, in general, dominate early colonization of tree leaves, gradually giving way to bacteria as decay advances (Triska, 1970; Kaushik and Hynes, 1971; Suberkropp and Klug, 1976; Barlocher and Kendrick, 1974). Chemical and structural modifications of the leaf material due to enzymatic and mechanical activity of microorganisms, such as bacteria and fungi are other conditioning mechanism (Gessner, 1999). Relative importance of the two mechanisms varies in relation to the composition of shredder taxa, fungal assemblage on leaves, extent of fungal colonization and the environmental conditions (Barlocher, 1985; Suberkropp, 1992a, 1998; Garca, 1993).

Catabolism

Catabolism is the biochemical term that describes energy yielding enzymatic reactions, or chains of reactions, usually involving the transformation of complex organic compounds to smaller and simpler molecules (Swift *et al.*, 1979). However, over a given time period, the catabolism of a given substrate or mixture of substrates may be incomplete. The products of catabolism include, inorganic (CO_2, NH_4+, PO_4^{-3} etc.) and other intermediate chemical constituents that enter the metabolic pool of the decomposer organisms which use them for resynthesis into the polysaccharides; others may be incorporated in non-cellular organic matter as humus (Swift *et al.*, 1979). Matter is lost from original resource by catabolic formation of volatile or soluble inorganic forms or by creation of soluble organic intermediates which are subsequently leached out. Over a relatively short period of time the original resource is acted upon by decomposer organisms to comminute and catabolise it to inorganic products, decomposer tissues, humus and comminuted particles of chemically unchanged residue. Over a much longer period decomposition may be complete and this time scale may be hundreds or even thousands of years because of the very slow turnover of the humus residues. Rather than being separate, leaching, conditioning and fragmentation to a large extent act simultaneously, allowing great potential for interaction. According to some workers (Meyer and O'Hop, 1983; Ward *et al.*, 1994), both microorganisms and shredders work mechanically and enzymatically on leaf tissue to affect patterns of solute and FPOM release. Catabolic activity softens plant material and renders it comminutable more readily; catabolic processes also result in the release of soluble components which may be removed in leachate; reduction in particle size improves the access of catabolic enzymes and increases the ease with which soluble compounds may be leached (Swift *et al.*, 1979).

Decomposition models

Most studies of litter breakdown have used the negative exponential model developed by Jenny *et al.* (1949) and Olson (1963) which is based on the assumption that the decomposition rate is proportional to the amount of matter remaining. Application of the single-exponential decay model implies that the combination of factors driving decomposition, such as substrate quality remain constant. Measure of the average rate of decomposition is the fractional loss rate 'k' and it is the slope of regression of $\log_e (Wt_o / Wt_n)$ where, Wt_o is the initial dry weight of the material, Wt_n the dry weight after time t which could be expressed in days, months or years. Since k is an exponent that characterizes the decomposition rate of a particular material, it is termed as decomposition constant. Since large number of studies have used single negative exponential model to study decomposition, as such k values of a wide diversity of organisms or their parts have been computed (Table 3.1).

As stated earlier, the exponential model of decomposition implies a constant decomposition rate, which is only a rough approximation of the pattern of decline in litter mass with time because decomposition constant varies widely with substrate composition. Thus, the process is more accurately described by a curve which represents the time course of leaf litter decomposition showing the major chemical constituents (cell solubles, celluloses and hemicelluloses, lignin). Such a model recognises three major phases of litter decomposition. During the first phase, leaching of cell solubles is the predominant process. The second phase of decomposition occurs more slowly and involves a combination of fragmentation by soil animals, chemical alteration by soil microbes, and leaching of decay products from the litter (Chapin *et al.*, 2002). Decomposition during this second phase is often measured as mass loss from dead leaves (Aerts, 1997), roots (Berg *et al.*, 1998) or twigs that are placed in mesh litter bags and weighed periodically (Vogt *et al.*, 1986; Robertson and Paul, 2000). The exponential model of decomposition has been applied primarily to this second phase (Chapin *et al.*, 2002). The final phase of decomposition occurs quite slowly and involves the chemical alteration of organic matter that is mixed with mineral soil and the leaching of breakdown products to other soil layers. Decomposition during this final phase is often estimated from measurements of soil respiration or isotopic tracers (Schlesinger, 1977; Trumbore and Harden, 1997). The decomposition rate and decomposition constant (k) gradually decline through these three phasesof decomposition.

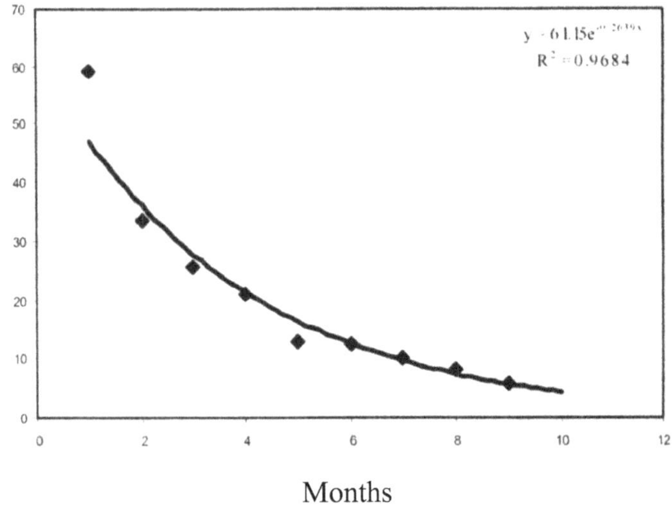

Fig.3.2. Fitting single negative exponential model to litter decomposition data. (**Source**: Tyub, 2004).

In view of the limitations of single negative exponential model, several alternative models which, however, are largely modifications of negative exponential model have been put forth (Table 3.2). For instance, Minderman (1968) proposed a model that accounts for the complex chemical nature of vascular plant material. In this model each class of chemicals breakdown at a constant rate and the overall rate of breakdown is the sum of the individual breakdown rates. Boiling *et al.*(1975) and Carpenter (1981) used a similar approach in modelling detritus dynamics. They divided organic material into compartments based on particle size and the extent of breakdown involving exchange of material among the compartments. A simple modification of the negative exponential model groups leaf components into two classes, labile and refractory. This double exponential model has been useful in describing breakdown in grasslands (Hunt, 1977: Wieder and Lang, 1982), shallow lakes (Brock et al., 1985) and laboratory simulation of leaf breakdown in tree boles (Carpenter, 1982).

Table 3.1. Decomposition constants of different species in aquatic and terrestrial ecosystems.

Plant type	Species	Fraction	Conditions	$k(d^{-1})$	Author
Aquatic plants					
Phytoplankton	Mixed natural community	Whole organism	Water	0.0360	Aizaki & Takamura, 1991
	Mixed natural community	Whole organism	Water	0.0560	Aizaki & Takamura, 1991
	Mixed natural community	Whole organism	Water	0.0660	Aizaki & Takamura, 1991
	Crysophyceae	Whole organism	Water	0.0360	Aizaki & Takamura, 1991
	Crysophyceae	Whole organism	Water	0.0360	Aizaki & Takamura, 1991
	Mixed natural community	Whole organism	Water	0.0470	Aizaki & Takamura, 1991
	Mixed natural community	Whole organism	Water	0.0680	Aizaki & Takamura, 1991
	Mixed natural community	Whole organism	Water	0.0270	Aizaki & Takamura, 1991
	Mixed natural community	Whole organism	Water	0.0980	Aizaki & Takamura, 1991
	Anabaena sp.	Whole organism	Water	0.0980	Aizaki & Takamura, 1991
	Microcystis sp.	Whole organism	Water	0.0528	Aizaki & Takamura, 1991
	Synechococcus sp.	Whole organism	Water	0.0956	Biddanda, 1988
	Dunaliana sp.	Whole organism	Water	0.0498	Biddanda, 1988
	Cylindrotheca sp.	Whole organism	Water	0.0417	Biddanda, 1988
	Seston	Whole organism	Water	0.0294	Garber, 1984
	Skeletonema costatum	Whole organism	Water	0.0449	Garber, 1984

Contd...

	Scenedesmus sp.	Whole organism	Water	0.0233	Golterman, 1972
	Skeletonema costatum	Whole organism	Water	0.0676	Newell *et al.*,1981
	Chaetoceros tricomutum	Whole organism	Water	0.0699	Newell *et al.*,1981
	Thalassiosira angstii	Whole organism	Water	0.0388	Newell *et al.*,1981
	Mixed natural community	Whole organism	Water	0.0093	Otsuki & Hanya, 1972
	Chloralla sp.	Whole organism	Water	0.0540	Twilley *et al.*,1986
	Mixed natural community	Whole organism	Water	0.0658	Walsh *et al.*, 1988
Macroalgae	*Macrocystis integrifolia*	Stipes	Water	0.0295	Albright *et al.*, 1980
	Macrocystis integrifolia	Photosynthetic tissue	Water	0.0321	Albright *et al.*, 1980
	Cladophora aft albida	Photosynthetic tissue	Water	0.0076	Albright *et al.*, 1983
	Cladophora aft albida	Photosynthetic tissue	Water	0.0082	Birch *et al.*, 1983
	Cladophora aft albida	Photosynthetic tissue	Water	0.0038	Gabrielson *et al.*, 1983
	Cladophora aft albida	Photosynthetic tissue	Water	0.0076	Gabrielson *et al*, 1983
	Cladophora aft albida	Photosynthetic tissue	Water	0.0035	Gabrielson *et al*, 1983
	Ulva sp.	Photosynthetic tissue	Water	0.0230	Twilley *et al.*, 1986

Contd...

Seagrasses	*Zostera marina*	Mixed litter	Water	0.0010	Godshalk & Wetze1.1978a
	Zostera marina	Mixed litter	Water	0.0020	Godshalk & Wetze1.1978a
	Zostera marina	Leaves	Water	0.0035	Harrison, 1982
	Zostera marina	Leaves	Water	0.0180	Harrison, 1982
	Thalassia testudinum	Leaves	Water	0.0007	Harrison, 1982
	Zostera marina	Leaves	Water	0.0070	Harrison, 1982
	Thalassia testudinum	Leaves	Water	0.0170	Harrison, 1982
	Thalassia testudinum	Leaves	Water	0.0085	Harrison, 1982
	Thalassia testudinum	Leaves	Water	0.0080	Harrison, 1982
	Posidonia australis	Leaves	Water	0.0013	Harrison, 1982
	Heterozostera tasmanica	Leaves	Water	0.0040	Harrison, 1982
	Zostera marina	Leaves (average)	Water	0.0124	Hemminga & Nieuwenhu-ize, 1991
	Cymodocea nodosa	Leaves (average)	Water	0.0230	Hemminga & Nieuwenhu-ize, 1991
	Thalassia testudinum	Rhizomes	Water	0.0007	Kenworthy & Thayer,1984
	Thalassia testudinum	Roots	Water	0.0183	Kenworthy & Thayer,1984
	Zostera marina	Roots	Water	0.0048	Kenworthy & Thayer,1984
	Zostera marina	Rhizomes	Water	0.0035	Kenworthy & Thayer,1984
	Thalassia testudinum	Leaves (average)	Water	0.0048	Newell *et al.*,1986

Contd...

	Thalassia testudinum	Leaves (average)	Water	0.0279	Newell *et al*.,1986
	Cymodocea nodosa	Leaves	Water	0.0039	Peduzzi & Herndl, 1991
	Zostera marina	Leaves	Water	0.0136	Pellikaan, 1982
	Zostera marina	Leaves	Water	0.0357	Pellikaan, 1984
	Zostera marina	Mixed litter	Water	0.0357	Pellikaan, 1984
	Posidonia oceanica	Mixed litter (+wood)	Water (20m)	0.0087	Romero *et al*,1992
	Posidonia oceanica	Mixed litter (+wood)	Water (5m)	0.0066	Romero *et al*,1992
	Thalassia testudinum	Leaves	Water	0.0149	Rublee & Roman, 1982
	Halophila stipulacea	Leaves	Water	0.0032	Wahbeh & Mahasneh, 1985
Fresh water angiosperms	*Potamogeton perfoliatus*	Leaves	Water	0.0446	Bastardo, 1979
	Potamogeton lucens	Leaves	Water	0.0517	Bastardo, 1979
	Potamogeton lucens	Leaves	Water	0.0458	Bastardo, 1979
	Elodea canadensis	Leaves	Water	0.0475	Bastardo, 1979
	Elodea canadensis	Leaves	Water	0.0859	Bastardo, 1979
	Ceratophyllum sp.	Leaves (average)	Water	0.0247	Best *et al*., 1990
	Vallisneria spiralis	Leaves	Water	0.0987	Briggs *et al*., 1985
	Najas flexilis	Leaves	Water	0.0070	Godshatk & Wetzel,1978a
	Myriophyllum heterophyllum	Leaves	Water	0.0090	Godshatk & Wetzel,1978a

Contd...

	Myriophyllum heterophyllum	Leaves	Water	0.0340	Godshalk & Wetze1,1978a
	Najas flexilis	Leaves	Water	0.0280	Godshalk & Wetze1,1978a
	Potamogeton nodosus	Leaves (average)	Water	0.0483	Hill, 1979
	Potamogeton crispus	Leaves	Water	0.0648	Rogers & Breen, 1982
	Potamogeton crispus	Leaves	Water	0.0640	Rogers & Breen, 1982
	Justicia americana	Leaves, petioles, stems	Water	0.0138	Twilley *et al.*,1985
	Justicia americana	Roots and rhizomes	Water	0.0398	Twilley *et al.*,1985
	Potamogeton sp.	Leaves	Water	0.0310	Twilley *et al.*,1985
	Ruppia sp.	Leaves	Water	0.0280	Twilley *et al.*,1985
	Myriophyllum sp.	Leaves	Water	0.0450	Twilley *et al.*,1985
Amphibious plants	*Sagittaria lancifolia*	Leaves	Water	0.0058	Twilley *et al.*,1985
	Sagittaria lancifolia	Stems	Water	0.0076	Twilley *et al.*,1985
	Nymphoides peltata	Petioles	Water	0.0420	Brock, 1984
	Nymphoides peltata	Long shoots	Water	0.0440	Brock, 1984
	Nymphoides peltata	Leaves	Water	0.0560	Brock, 1984
	Nymphoides peltata	Leaves	Water	0.0910	Brock, 1984
	Nymphoides peltata	Petioles	Water	0.0450	Brock, 1984
	Nymphoides peltata	Roots	Water	0.0790	Brock, 1984

Contd...

	Nymphoides peltata	Roots	Water	0.0490	Brock, 1984
	Nymphoides peltata	Short shoots	Water	0.0350	Brock, 1984
	Nymphoides peltata	Long shoots	Water	0.0370	Brock, 1984
	Nymphoides peltata	Short shoots	Water	0.0550	Brock, 1984
	Nuphar variegatum	Leaves	Water	0.0600	Godshalk & Wetzel,1978a
	Nuphar variegatum	Leaves	Water	0.0200	Godshalk & Wetzel,1978a
	Sparganium eurycarpum	Mixed litter	Water	0.0076	Neeley & Davis, 1985
	Sparganium eurycarpum	Mixed litter	Water	0.0021	Neeley & Davis, 1985
	Sparganium eurycarpum	Mixed litter	Water	0.0017	Neeley & Davis, 1985
	Eichhornia crassipes	Mixed litter (average)	Water	0.0095	Reddy & DeBusk, 1991
	Nuphar luteum	Leaves, petioles, stems	Water	0.0988	Twilley *et al*,1985
	Nuphar luteum	Roots and rhizomes	Water	0.0142	Twilley *et al*,1985
Terrestrial plants					
Sedges	*Phragmites communis*	Mixed litter	Water	0.0018	Andersen, 1978
	Phragmites communis	Mixed litter	Water	0.0014	Andersen, 1978
	Panicum sp.	Mixed litter	Water	0.0071	Bayley *et al.*, 1985
	Spartina altemiflora	Roots	Below-ground	0.0067	Benner *et al.*,1991

Contd...

	Spartina alternifolia (short form)	Mixed litter	Soil/ Fertilized	0.0052	Breteler & Teal, 1981
	Spartina alternifolia (tall form)	Mixed litter	Soil/ Fertilized	0.0081	Breteler & Teal, 1981
	Spartina alternifolia (short form)	Mixed litter	Soil/Control	0.0033	Breteler & Teal, 1981
	Spartina alternifolia (short form)	Mixed litter	Soil/Control	0.0063	Breteler & Teal, 1981
	Typha domingensis	Mixed litter	Water	0.0010	Davis, 1991
	Typha domingensis	Mixed litter	Water	0.0099	Davis, 1991
	Cladium jamaicense	Mixed litter	Water	0.0013	Davis, 1991
	Cladium jamaicense	Mixed litter	Water	0.0007	Davis, 1991
	Cladium jamaicense	Mixed litter	Water	0.0007	Davis, 1991
	Typha domingensis	Mixed litter	Water	0.0021	Davis, 1991
	Typha marsh	Mixed litter	Water	0.0010	Findely *et al.*, 1990
	Scirpus subterminalis	Mixed litter	Water	0.0090	Godshalk & Wetzel,1978a
	Scirpus acutus	Mixed litter	Water	0.0020	Godshalk & Wetzel,1978a
	Scirpus acutus	Mixed litter	Water	0.0050	Godshalk & Wetzel,1978a
	Scirpus subterminalis	Mixed litter	Water	0.0020	Godshalk & Wetzel,1978a
	Spartina altemiflora	Mixed litter	Water	0.0111	Haines & Hanson, 1979
	Juncus roemerianus	Mixed litter	Water	0.0091	Haines & Hanson, 1979

Contd...

	Spartina anglica	Mixed litter	Water	0.0079	Hemminga & 13011, 1991
	Spartina anglica	Mixed litter	Water	0.0022	Hemminga & 13011, 1991
	Triglochin maritinta	Leaves	Water	0.0256	Hemminga & 13011, 1991
	Spartina anglica	Mixed litter	Water	0.0033	Hemminga & 13011, 1991
	Spartina anglica	Mixed litter	Water	0.0093	Hemminga & 13011, 1991
	Triglochin maritinta	Mixed litter	Water	0.0025	Hemminga & 13011, 1991
	Spartina anglica	Leaves	Water	0.0061	Hemminga & 13011, 1991
	Typha glauca	Mixed litter	Water	0.0011	Neeley & Davis, 1985
	Typha glauca	Mixed litter	Water	0.0016	Neeley & Davis, 1985
	Typha glauca	Mixed litter	Water	0.0011	Neeley & Davis, 1985
	Typha glauca	Leaves (senesced)	Water	0.0104	Nelson *et al*., 1990
	Typha glauca	Leaves (green)	Water	0.0235	Nelson *et al*., 1990
	Juncus roemerianus	Mixed litter	Water	0.0017	Newell *et al*.,1984
	Phragmites communis	Leaves	Water	0.0045	Tanaka, 1991
	Spartina sp.	Mixed litter	Water	0.0098	Twilley *et al*.,1986
	Spartina alternifolia	Mixed litter	Water	0.0043	Valiela *et al*., 1984
	Spartina alternifolia	Mixed litter	Water	0.0071	Valiela *et al*., 1984
	Typha glauca	Mixed litter	Water	0.0012	van der Valk *et al*., 1991
	Scolochloa festucacea	Mixed litter	Water	0.0016	van der Valk *et al*., 1991
	Scirpus lacustris	Mixed litter	Water	0.0010	van der Valk *et al*., 1991

Contd...

	Phragmites australis	Mixed litter	Water	0.0007	van der Valk *et al.*, 1991
	Scolochloa festucacea	Mixed litter	Water	0.0022	van der Valk *et al.*, 1991
	Typha x glauca	Mixed litter	Water	0.0012	van der Valk *et al.*, 1991
	Phragmites australis	Mixed litter	Water	0.0003	van der Valk *et al.*, 1991
	Scirpus lacustris	Mixed litter	Water	0.0015	van der Valk *et al.*, 1991
	Typha x glauca	Mixed litter	Water	0.0010	van der Valk *et al.*, 1991
	Scolochloa festucacea	Mixed litter	Water	0.0023	van der Valk *et al.*, 1991
	Phragmites australis	Mixed litter	Water	0.0008	van der Valk *et al.*, 1991
	Scirpus lacustris	Mixed litter	Water	0.0011	van der Valk *et al.*, 1991
Mangroves	*Kandelia candel*	Mixed litter (+wood)	Water	0.0018	Lee, 1989
	Avicennia marina				
	Avicennia marina	Mixed litter (+wood)	Water	0.0095	Newell *at al.*,1984
	Rhizophora spp.	Mixed litter (+wood)	Soil	0.0008	Robertson & Daniel,1989
	Rhizophora spp.	Mixed litter (+wood)	Soil	0.0002	Robertson & Daniel,1989
	Avicennia marina	Mixed litter (+wood)	Water/ bagged	0.0114	van der Valk & Attiwill, 1984
	Avicennia marina	Mixed litter (+wood)	Water/ unbagged	0.0189	van der Valk & Attiwill, 1984

Contd...

	Avicennia marina	Roots	Water	0.0038	van der Valk & Attiwill, 1984
	Avicennia marina	Leaves	Water	0.0071	van der Valk & Attiwill, 1984
Grasses	*Molinia caerulea*	Mixed litter	Soil	0.0153	Aerts, 1989
	Elymus pycnanthus	Mixed litter	Water	0.0079	Hemminga & Buth, 1991
	Etythrina sp.	Mixed litter	Soil	0.0095	Palm & Sanchez, 1990
	Cajanus cajan	Mixed litter	Soil	0.0047	Palm & Sanchez, 1990
	Inge edulis	Mixed litter	Soil	0.0025	Palm & Sanchez, 1990
	Tallgrass prairie	Mixed litter (average)	Soil	0.0009	Seastedt, 1988
Broad deciduous tree leaves	*Red maple*	Leaves	Soil	0.0020	Aber *et al.*, 1990
	Red oak	Leaves	Soil	0.0011	Aber *et al.*, 1990
	Aspen	Leaves	Soil	0.0014	Aber *et al.*, 1990
	Red oak	Leaves	Soil	0.0011	Aber *et al.*, 1990
	Sugar maple	Leaves	Soil	0.0023	Aber *et al.*, 1990
	Paper birch	Leaves	Soil	0.0017	Aber *et al.*, 1990
	Red maple	Leaves	Soil	0.0019	Aber *et al.*, 1990
	Red oak	Leaves	Soil	0.0009	Aber *et al.*, 1990
	White oak	Leaves	Soil	0.0012	Aber *et al.*, 1990
	Sugar maple	Roots	Soil	0.0006	Aber *et al.*, 1990
	Alnus incana	Leaves	Soil	0.0009	Berg & Ekbohm, 1991

Contd...

	Betula pubescens	Leaves	Soil	0.0009	Berg & Ekbohm, 1991
	Betula pubescens	Leaves	Soil	0.0009	Berg & Ekbohm, 1991
	Populus tremuloides	Leaves	Soil	0.0012	Bockheim *et at*, 1991
	Quercus eslllpsoidalis	Leaves	Soil	0.0009	Bockheim *et at*, 1991
	Betula papyrifera	Leaves	Soil	0.0012	Bockheim *et at*, 1991
	Frangula alnus	Leaves	Soil	0.0054	Escudero *et at*.,1991
	Quercus pyrenaica	Leaves	Soil	0.0030	Escudero *et at*.,1991
	Betula pubescens	Leaves	Soil	0.0033	Escudero *et at*.,1991
	Salix fragilis	Leaves	Water	0.0246	Gessner *et at*., 1991
	Alnus glutinosa	Leaves	Soil	0.0252	Gessner *et at*., 1991
	Fagus sylvatica	Leaves	Soil	0.0007	Gosz *et at*,1973
	Sugar maple	Leaves	Soil	0.0014	Gosz *et at*,1973
	Sugar maple	Leaves	Soil	0.0009	Gosz *et at*,1973
	Yellow birch	Leaves	Soil	0.0017	Gosz *et at*,1973
	Yellow birch	Leaves	Soil	0.0023	Gosz *et at*,1973
	Fagus sylvatica	Leaves	Soil	0.0010	Gosz *et at*,1973
	Fagus sylvatica	Leaves	Water	0.0035	Iversen, 1973
	Fagus sylvatica	Leaves (average)	Soil	0.0021	Joergensen & Meyer, 1990
	Fagus sylvatica	Leaves	Soil	0.0013	Joergensen, 1991
	Aspen	Leaves	Soil	0.0016	McClaugherty *et at*., 1985

Contd...

	White oak	Leaves	Soil	0.0015	McClaugherty *et al*, 1985
	Red maple	Wood chips	Soil	0.0008	McClaugherty *et al*, 1985
	Sugar maple	Leaves	Soil	0.0022	McClaugherty *et al*, 1985
	Alnus nepalensis	Wood part	Soil	0.0029	Sharma & Ambasht, 1987
	Aspen	Leaves	Soil	0.0018	Taylor *et al*,1989
	Balsam poplar	Leaves	Soil	0.0016	Taylor *et al*,1989
	Cow-parsnip	Mixed litter	Soil	0.0036	Taylor *et al*,1989
	Grass	Mixed litter	Soil	0.0022	Taylor *et al*,1989
	Dogwood leaf litter	Leaves	Soil	0.0021	Taylor *et al*,1989
Shrubs	*Salicornia virginica*	Mixed litter (+wood)	Water	0.0413	Haines & Hanson, 1979
	Halimione portulacoides	Mixed litter (+wood)	Water	0.0090	Hemminga & Buth, 1991a
	Limonium vulgare	Mixed litter (+wood)	Water	0.0025	Hemminga & Buth, 1991a
	Limonium vulgare	Mixed litter (+wood)	Water	0.0048	Hemminga & Buth, 1991a
	Halimione portulacoides	Mixed litter (+wood)	Water	0.0090	Hemminga & Buth, 1991a
	Leucospermum parile	Mixed litter (+wood)	Soil	0.0002	Mitchell *et al*., 1986
	Acacia urophylla	Mixed litter (+wood)	Soil	0.0010	O'Conne11,1987

Contd...

	Trymalium spathulatum	Leaves	Soil	0.0031	O'Conne11,1987
	Bossiaea laidlawaiana	Leaves	Soil	0.0016	O'Conne11,1987
	B. laidlawaiana pods	Pods	Soil	0.0008	O'Conne11,1987
	Casuarina decussata	Leaves	Soil	0.0012	O'Conne11,1987
	Acacia urophylla	Leaves	Soil	0.0015	O'Conne11,1987
	Ceanothus megacarpus	Leaves	Soil	0.0010	Schlesinger, 1985
	Salvia melifera	Leaves	Soil	0.0011	Schlesinger, 1985
	Salvia melifera	Leaves	Soil	0.0009	Schlesinger, 1985
	Ceanothus megacarpus	Leaves	Soil	0.0010	Schlesinger, 1985
	Rosa sp.	Leaves	Soil	0.0032	Taylor *et al*,1989
	Mallotus philippensis	Leaves	Soil	0.0110	Upadhyay *et al.*,1989
Conifers	*Pinus contorta*	Needles	Soil	0.0004	Yavitt & Fahey, 1986
	White pine	Roots	Soil	0.0008	Aber *et al.*, 1990
	Hemlock	Needles	Soil	0.0010	Aber *et al.*, 1990
	White pine	Needles	Soil	0.0010	Aber *et al.*, 1990
	Red pine	Needles	Soil	0.0009	Aber *et al.*, 1990
	Scots pine	Needles	Soil	0.0008	Berg *et al.*, 1982
	Scots pine	Needles	Soil	0.0007	Berg *et al.*, 1982
	Scots pine	Needles	Soil	0.0009	Berg *et al.*, 1982
	Pinus sylvestris	Needles	Soil	0.0008	Berg & Ekbohm, 1991

Contd...

	Pinus sylvestris	Needles	Soil	0.0010	Berg & Ekbohm, 1991
	Lodgepole pine	Needles	Soil	0.0008	Berg & Ekbohm, 1991
	Lodgepole pine	Needles	Soil	0.0008	Berg & Ekbohm, 1991
	Brown spruce	Needles	Soil/ Fertilized	0.0006	Berg & Ekbohm, 1991
	Brown spruce	Needles	Soil	0.0005	Berg & Tamm, 1991
	Green spruce	Needles	Soil/ Fertilized	0.0008	Berg & Tamm, 1991
	Green spruce	Needles	Soil	0.0010	Berg & Tamm, 1991
	Pinus banksiana	Needles	Soil	0.0005	Bockheim *et al*., 1991
	Pinus pinaster	Needles	Soil	0.0010	Bockheim *et al*., 1991
	Pinus sylvestris	Needles	Soil	0.0020	Bockheim *et al*., 1991
	Sitka spruce	Branches	Soil	0.0355	Fahey *et al*., 1991
	Quercus lanuginose	Leaves	Soil	0.0049	Upadhyay *et al*., 1989
	Lyonia ovalifolia	Leaves	Soil	0.0073	Upadhyay *et al*., 1989
	Quercus glauca	Leaves	Soil	0.0073	Upadhyay *et al*., 1989
	Shorea robusta	Leaves	Soil	0.0076	Upadhyay *et al*., 1989
	Quercus floribunda	Leaves	Soil	0.0051	Upadhyay *et al*., 1989
	Quercus leucotrichophora	Leaves	Soil	0.0052	Upadhyay *et al*., 1989

Contd...

Broad perennial tree leaves	*Eucalyptus diversicolor*	Fruit	Soil	0.0005	O'Connell,1988
	Eucalyptus diversicolor	Leaves	Soil	0.0015	O'Connell,1988
	Eucalyptus diversicolor	Twigs	Soil	0.0003	O'Connell,1988
	Eucalyptus diversicolor	Bark	Soil	0.0006	O'Connell,1988
	Myrica esculenta	Leaves	Soil	0.0043	Upadhyay *et al.*, 1989
	Rhododendron arboreum	Leaves	Soil	0.0048	Upadhyay *et al.*, 1989

Source : Enriquery *et al.* (1993)

Table 3.2. Commonly used decomposition models.

Model	Expression	References
Single-exponential model	$X=X_0e^{-kt}$	Jenny *et al*, 1949, Olson, 1963, Wieder and Lang, 1982, Harmon *et al*, 1986
Multiple-exponential model	$X=X_{0,1}\, e^{-k_1 t} + X=X\, e^{-k_2 t} + X=X_{0,3} e^{-k_3 t}$	Wieder and Lang, 1982, Means *et al.*, 1985, Minderman, 1968
Lag-time model	$X= 1 - (1 - exp[-kt])^N$	Harmon *et al*, 1986
Linear model	$X= X_0 - kt$	Wieder and Lang, 1982, Lambert *et al*, 1980

Another model which is similar to the double exponential model is termed as asymptotic model in which refractory material is assumed to be completely resistant to decay, hence remains unchanged while whole labile fractions disappear (Force and McCarty, 1970; Jewell, 1971).

Minshall and Minshall (1978) and Short *et al.* (1984) also modified negative exponential model in various ways to account for the effects of temperature on breakdown rates. A more complex modification of negative exponential model is that of Hanson *et al.*, (1984) in which breakdown rate is a linear function of temperature. In short simulation models for detrital breakdown have incorporated temperature effects in various ways. Webster (1983) used the degree-day model in simulating organic matter dynamics in a second order stream and Boiling *et al.* (1975) used a Q_{10} function to modify breakdown rates with respect to temperature. Some workers (Bunell *et al.*, 1977a, b; Carpenter and Adams, 1979) also used empirically determined relationships between breakdown rates and temperature. Saunders (1976) modified negative exponential model to include availability of microbial enzymes necessary for decomposition to proceed. Parnas (1975) developed a model in which decomposition rate is entirely based on the growth rate of microbial organisms (see Chapter 5).

Models incorporating specific features of litter decomposition in aquatic ecosystems have also been put forth. For example, dynamics of benthic organic matter in a stream dominated by allochthonous input has been modelled by Fisher and Likens (1973) as:

$$\frac{dX}{dt} = I - kX - k'X$$

where X is the standing crop of benthic organic matter (mass or energy per unit area) and dX /dt is the rate of change of X through time t, I is input of particulate material (litter fall, mass or energy per area per unit time), and k is the exponent in the negative exponential model, which has been commonly applied to particulate matter in streams (Peterson and Cummins, 1974). The term k'X is the rate at which the material is being transported downstream i.e., the exponential loss rate caused by transport. An analogous model for particle transport is dX/dx = kxX, where X is the particulate material in transport, dX/dx is the change of X over distance and kx is the exponential rate at which material is lost from the water column to the bottom per unit distance.

kx can be measured by releasing particles into the water column and measuring water column concentrations downstream. The inverse of kx is the average transport distance and is designated as S_x i.e. S_x =kx/ Sx. So, Sx is the average distance a particle is transported in the water column before it hits the bottom and is retained. This model has been used for drifting insects (McLay, 1970; Elliott, 1971; Lancaster *et al.*, 1996), FPOM (Webster *et al.*, 1987; Miller and Georgian, 1992; Cushing *et al.*, 1993), leaves (Speaker *et al.*, 1984; Speaker *et al.*, 1988; Cummins *et al.*, 1989) and wood (Trotter, 1990; Covich and Crowl, 1990; Ehrman and Lamberti, 1992; Webster *et al.*, 1994).

With respect to spatial pattern, decomposition is heterogeneous at several scales. Most decomposition occurs near the soil surface, where litter inputs are concentrated. Soil mixing by animals, especially termites and earthworms, and leaching of dissolved organic matter also transfer surface carbon to varying depths (Reardon and Forbes, 2001). About half of the soil organic carbon, therefore, is typically below 20 cm depth, even though only a third of the roots are below that depth (Jobbagy and Jackson, 2000). The deep soil carbon is often older, more recalcitrant, and more strongly protected by its complexation with soil minerals than is surface carbon (Trumbore and Harden, 1997).

3.2. Patterns

Litter decomposition

Decomposition studies have focussed primarily on litter of single species and such studies have helped in elucidating the role of climate, quality of litter and associated decomposer community on litter decomposition. However, very recently a new dimension of potential interactions among litters of different species during decomposition has been shown to be significant (Gartner and Cardon, 2004). In fact, Seastedt (1984) suggested that due to differences in resource quality between species, litter mixtures might decompose at a different rate to that which would be predicted from single species decomposition studies. Understanding of interactions between littors of different species is essential because litter does not segregate into individual species types in ecosystems and the composition of plant communities also changes over time. Characteristics of decomposition in litter mixtures that deviate from responses predicted from decomposition of single species litters are designated as "non-additive". Responses in mixtures when predictable from decay of component species alone, are referred to as "additive patterns". Comparative decomposition studies of mixed and single species litters reviewed by Gartner and Cardon (2004) clearly reveal that the decomposition patterns are not always predictable from single-species dynamics and non-additive patterns of mass loss are more prevalent. Among the studies reviewed by Gartner and Cardon (2004) 67% of the studies exhibited non-additive mass loss and more often, mass loss in mixtures exceeded expected decay by only 20% or less. This pattern is irrespective of the nature of the species in the litter mixture. Synergistic (non-additive) effect in mass loss has been obtained whether different but only broadleaved species were mixed (Montagnini *et al.*, 1993; Briones and Meson, 1996), broadleaved species were mixed with needle litter (Rustad and Cronan, 1988; Conn and Dighten, 2000), litter of different grasses (Hector *et al.*, 2000) or of multiple functional groups (Robinson *et al.*, 1999) was mixed. Such litter decomposition patterns have been attributed to variety of factors but litter traits, in particular, are of major importance. For instance, higher nitrogen content of the litter mixture has a synergistic effect on decomposition (Wardle *et al.*, 1997). Contrastingly, studies have also shown slower decay rates than expected when litter of different species is mixed. For example, inclusion of *Empetrum hermaphroditum* litter in the boreal forest litters retards mass loss (Nilsson *et al.*, 1998). Inhibition of microbial activity by secondary metabolites released by E. *hermaphroditum* is known cause of such decline in the decay rate of litter when in mixture. The additive and non-additive patterns observed in litter mixtures also show inter-annual variations. Hansen and Coleman (1998) observed 7% greater mass loss in field litter bags than predicted in mixtures of three deciduous tree species after nine months of decay. In the next year 9% less mass loss was observed after 10 months in second cohort of field litter bags of same three species in the same forest. After 18 months of decay, litter mass loss did not differ significantly from expected values (calculated from single species loss) for either cohort. However, it needs to be emphasized that litter decomposition studies, in future, need to consider this hitherto unknown aspect also with emphasis on identifying the key species and key traits responsible for synergistic or antagonistic interactions between decomposing litter of different species.

Root decomposition

Decomposition of roots is responsible for significant carbon flux in terrestrial ecosystems, particularly with high below-ground allocation, such as grasslands (Seastedt, 1984) or some tropical forests (Vogt *et al.*, 1986; Nepstad *et al.*, 1994; Silver *et al.*, 2000). Root decay is also an important source of mineral nutrients in the soil (Persson, 1978; Fogel and Hunt, 1979; Aerts *et al.*, 1992) and thus influences rate of net primary productivity in nutrient limited environments. Notwithstanding this importance, surprisingly little attention has been paid to the study of root decomposition. Since the roots are exposed to a different decomposition environment than above-ground tissues, principles of decomposition established for litter decomposition may not hold true for root decomposition. However, recently Silver and Miya (2001), using a global data set, attempted to establish general patterns and principles of root decomposition. Like litter decomposition, both exponential and linear decay models have been used to describe mass loss of roots over a period of time (Aber *et al.*, 1990; Gijsman *et al.*, 1997). Some models incorporate a two stage root decomposition process (Fogel and Hunt, 1979; McClaugherty *et al.*, 1984). First stage is characterized by rapid mass loss driven primarily by inorganic chemical composition and water soluble carbon loss via microbial utilization or leaching while the second stage is characterized by slower mass loss regulated largely by lignin and other recalcitrant root materials (McClaugherty *et al.*, 1984).

Comparison of root and litter decomposition during the **Long Term Intersite Decomposition Experiment (LIDET)** over a five year period in 28 sites in North and central America (Gholz *et al.*, 2000) revealed that the ratio of above- to below-ground decomposition rate is 1.63 in tropical wet forests and 0.87 in tropical dry forests, indicating that leaf litter decomposes faster than roots in wet forests while roots decompose faster than leaves in dry forests. In addition, substrate quality also influences root decomposition as in leaf litter but its role in decay is different (see Chapter 4 for detailed discussion). Concentration of mineral nutrients and secondary metabolites, root diameter and ratios of carbon: nutrient and nutrient: nutrient are also responsible for determining root litter quality and its decomposition(Boot, 1990).

Limit values for decomposition

Decomposition of litter is accompanied by formation of soil organic matter (SOM) and release of nutrients. However, part of it accumulates because of being recalcitrant or extremely slow to decomposition (Osono and Takeda, 2005). For instance, decomposition rate of the stable fraction of Scots pine needle litter is about 1% in 30-300 years (Couteaux *et al.*, 1998). Size of such a slowly decomposing fraction has been estimated making use of the limit value for accumulated mass loss (Berg *et al.*, 1995b; 1996) which is calculated as an asymptotic value towards which the decomposition proceeds using the non-linear model of Howard and Howard (1974) and Berg and Ekbohm (1991):

$$m.1 = m(1-e^{-kt/m})$$

where m.1 is the accumulated mass loss (in percent) and t is time (days, months, years). The parameter m represents the maximum accumulated mass loss (asymptotic level), and the parameter k is the initial decomposition rate. Limit values were initially reported for litter types incubated in soil systems more or less free of soil animals (Howard and Howard, 1974) but such values are known for a range of species from habitats even with high animal activity (Table 3.3).

Lignocellulose index (LCI, the ratio of acid-soluble holocellulose to acid-soluble holocellulose plus acid-insoluble lignin and lignin-like substances) is a simple but useful descriptor of carbon fractions in decomposing litter and SOM (Berg *et al.*, 1984; Melillo *et al.*, 1989; Aber *et al.*, 1990). LCI in freshly shed leaves ranges from 0.3 to 0.8 among litter types in temperate and boreal forests (Table 3.4). It gradually decreases due to rapid decomposition of holocellulose than lignin and ultimately approaches asymptotic values when the disappearance of holocellulose and lignin proceeds at the same rate and decomposing litter becomes SOM. Numerous studies have shown strong correlation between the limit values for accumulated mass loss and their initial LCI and lignin content (Osono and Takeda, 2005). However, Berg *et al.* (1996), Berg and Johansson (1998) reported a weak relationship between limit values and LCI in temperate and boreal forests. Instead, these studies indicated stronger relationship between the limit values and nutrient contents, such as N, Ca and Mn. Polyphenols and K contents have also shown correlations with limit values but their

Table 3.3. Limit values (% of initial mass) for accumulated mass in different litter types from temperate and boreal forests.

Site	Species	Litter type	Limit value	Reference
Europe				
Jadraas, Sweden	*Pinus sylvestris*	C	89.0	Berg and Ekbohm, 1991
	Pinus contorta	C	100.0	
	Betula pubescens	B	56.9	
	Alnus incana	B	50.6	
Malung, Sweden	*Pinus sylvestris*	C	84.5	Berg and Ekbohm, 1993
	Pinus contorta	C	97.0	
Strasan, Sweden	*Picea abies*	C	61.0	Berg and Tamm, 1991
Monte Taburno, Italy	*Fagus grandifolia*	B	57.8	
	Abies alba	C	51.3	
Meathop, UK	*Corylus avellana*	B	40.3	Howard and Howard, 1974
	Quercus petraea,	B	43.1	
	Q. robur		42.2	
	Betula pendula	B	37.8	
	Tilia cordata	B	35.0	
	Fraxinus excelsior	B	65.0	
	Ulmus glabra	B	64.6	Anderson, 1973
Bleam woods, UK	*Fagus sylvatica*	B	80.0	
	Castanea sativa	B		
North America				
Wisconsin, USA	*Pinus strobus*	C	84.3	McClaugherty *et al.*,
	Quercus alba	B	88.9	1985

Contd...

Kananaskis, Canada	*Populus grandidentata* *Acer saccharum* *Quercus rubra* *Tsuga canadensis* *Populus tremuloides*	B B C B	80.0 83.0 85.4 70.0 56.4	Lousier and Parkinson, 1976
Japan Shiga Kyoto	*Pinus thunbergii* *Hydrangea hirta* *Clerodendrumttichoto-mum* *Deutzia crenata* *Mallotus japonicus* *Swida controversa* *Pterostyrax hispida* *Carpinus laxiflora* *Castanea crenata* *Magnolia obovata* *Quercus crispula* *Acer mono var. marmoraturn* *Betula grossa* *Acer rufinerve* *Cryptomeria japonica* *Pterocarya rhoifolia* *Fagus crenata* *Aesculus turbinata*	C B B B B B B B B B B B B B C B B B	75.6 74.6 96.6 90.1 90.7/92.2 94.1/71.8 79.1/78.8 78.0/75.9 68.8/73.3 73.1/80.9 76.7/72.9 57.1/57.8 69.7/76.4 59.1/59.5 79.4/58.5 49.3/58.6 56.7/52.9 53.0/46.8	Bhatta, 2003 Bhatta *et al.*, cf. Osono and Takeda, 2005 !Osono and Takeda, 2005

Source: OsonoandTakeda(2005);B=Broadleaved,C=Conifer;!Limitvaluesforuppersite(moder)/limitvaluesforlowersite(mull).

Table 3.4. Initial and final lignocellulose index (LCI) in different litter types.

Site	Species	Litter type	Study period (months)	LCI		Reference
				Initial	Final	
Europe						
Ivantjarnsheden, Sweden	*Pinus sylvestris*	C	36	0.67	0.50	Berg *et al.*, 1982
Ivantjarnsheden, Sweden	*Pinus sylvestris*	C	65	0.66	0.44	Berg *et al.*, 1982
	Vaccinium vitis-idae	B	24	0.52	0.42	
Ivantjarnsheden, Sweden	*Betula pubescen*	B	48	0.58	0.44	Berg and Wessen, 1984
Sweden	*Pinus sylvestris*	C	36	0.68	0.48	
Ivantjarnsheden, Sweden	*Pinus sylvestris*	C	48	0.65	0.44	Berg *et al.*, 1987
North America						
Wisconsin, USA	*Acer saccharum*	B	24	0.78	0.48	McClaugherty *et al*, 1985
	Populus grandidentata	B	24	0.69	0.42	
	Quercus alba	B	24	0.70	0.53	
	Pinus strobus	C	24	0.67	0.54	
	Tsuga canadensis	C	24	0.66	0.51	
	Quercus borealis	B	24	0.65	0.45	
Wisconsin, USA	*Abies amabilis*	C	72	0.49	0.37	Edmond, 1984
California, USA	*Ceanothus megacarpus*	B	36	0.49	0.27	Schlesinger, 1985
	Salvia mellifera	B	36	0.65	0.46	
	Pinus resinosa	C	77	0.55	0.33	Mellilo *et al.*, 1989
Massachusetts, USA	*Pinus resinosa*	C	72	0.60	0.39	
Massachusetts, USA	*Acer rubrum*	B	72	0.68	0.37	

Contd...

	Betula alleghaniensis	B	72	0.71	0.35	
	Quercus velutina	B	72	0.61	0.35	
Japan						
Shimane, Japan	*Quercus serrata*	B	36	0.38	0.36	Salamanca *et al*., 1998
Shiga, Japan	*Pinus thunbergii*	C	36	0.62	0.27	Bhatta, 2003
	Mallotus japonicus	B	18	0.57	0.24	Bhatta *et al*., cf. Osono and Takeda, 2005
	Pauownia tomentosa	B	18	0.39	0.25	
	Hydrangea hirta	B	18	0.53	0.27	
	Acanthopanax sciadophylloides	B	18	0.46	0.25	
	Clerodendrum tric-hotomum	B	18	0.45	0.33	
	Swida controversa	B	18	0.44	0.29	
	Pterostyrax hispida	B	18	0.40	0.27	
	Symplocos chinensis	B	18	0.39	0.25	
	Weigela hortensis	B	18	0.38	0.26	
Kyoto, Japan	*Deutzia crenata*	B	18	0.37	0.25	Osono and Takeda, 2005
	Mallotus japonicus	B	18	0.62	0.27/0.32	
	Swida controversa	B	36	0.57	0.24/0.19	
	Pterostyrax hispida	B	36	0.39	0.25/0.24	
	Carpinus laxiflora	B	36	0.53	0.27/0.32	
	Castanea crena	B	36	0.44	0.29/0.28	
	Magnolia obovata	B	36	0.46	0.25/0.26	
	Quercus crispula	B	36	0.40	0.27/0.29	
	Acermono var.. marmoratum	B	36	0.37	0.25/0.21	

Contd...

	Betula grossa	B	36	0.39	0.25/0.27	
	Acer rufinerve	B	36	0.26	0.21/0.20	
	Cryptomeria japonica	C	36	0.38	0.26/0.27	
	Pterocarya rhoifolia	B	36	0.28	0.20/0.20	
	Fagus crenata	B	36	0.46	0.33/0.34	
	Aesculus turbinata	B	36	0.27	0.21/0.20	

Source: Osono and Takeda (2005); B = Broadleaved, C = Conifer; I LCI values of species for upper site moder)/ LCI values of species for lower site (mull).

importance in the process of SOM formation remains obscure owing to their rapid leaching from fresh litter.

Melillo *et al.* (1989) and Aber *et al.* (1990) reported that LCI of different litter types in temperate regions reveals a convergent trend and approaches values of SOM in the underlying humus layer, while Berg *et al.* (1984) recorded a weak convergent trend among different litter types. This dichotomy of results merits further investigations across different ecosystems and litter types. In addition limit values also differ when compared across sites and ecosystems indicating that regulatory factors may vary in different ecosystems and sites (Table 3.4).

Decomposition pattern in ecosystems

In both aquatic and terrestrial ecosystems, transfer of fixed carbon to herbivores and decomposers/detritivores represents the major pathway of material flow. Cebrian and Lartigue (2004) compiled an extensive data set (350 reports with data for >800 systems) so as to unravel the patterns of herbivory and decomposition in aquatic and terrestrial ecosystems. Their study revealed that percentage of net primary production (NPP) channelled as detritus varied from ~0% to 100% within aquatic ecosystems, and from ~25% to 100% within terrestrial ecosystems. In most of the terrestrial and aquatic ecosystems, more than 50% of NET was processed through the detrital pathway. However, terrestrial ecosystems tend to channel a higher percentage of NPP as detritus compared to aquatic ecosystems. Detrital production was also higher in terrestrial ecosystems and also varied widely, particularly, within aquatic ecosystems. Decomposition rates (proportion of detrital mass decomposed per day) though varied widely within both aquatic as well as terrestrial ecosystems but were higher in aquatic systems. Absolute decomposition (g $C.m^{-2}.yr^{-1}$) also varied widely but did not differ significantly between the aquatic and terrestrial ecosystem types.

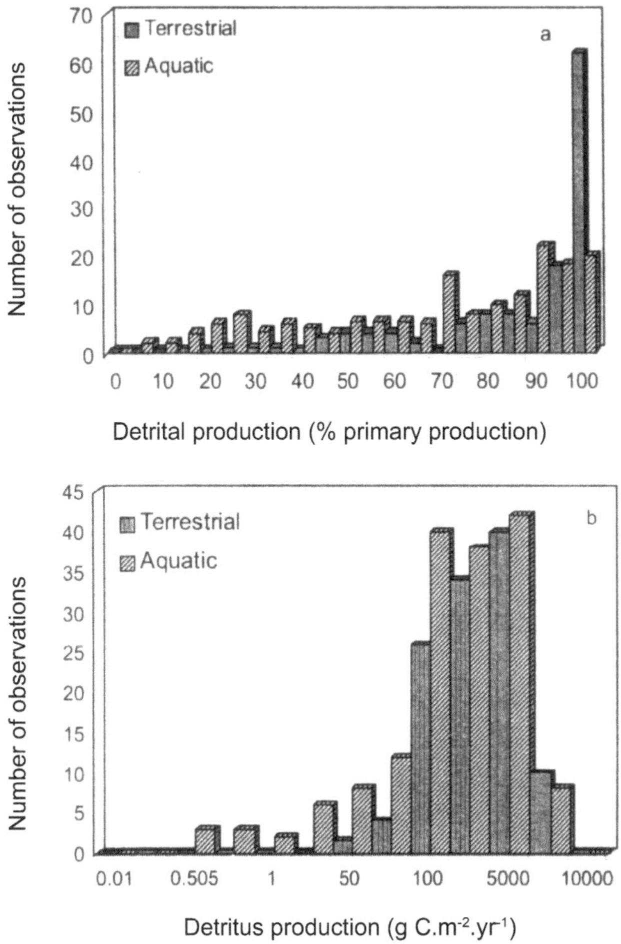

Fig. 3.3 Comparison of detrital production as percentage of primary production (a) and absolute detrital production (b) in terrestrial and aquatic ecosystems.

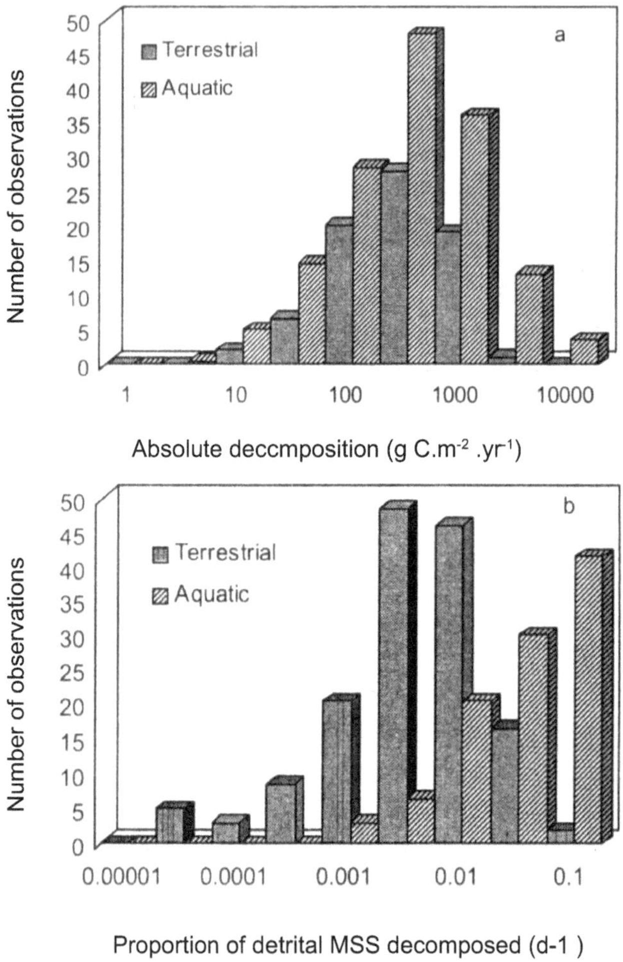

Fig. 3.4 Comparison of absolute decomposition (a) and proportion of detrital mass decomposed per day (b) in terrestrail and aquatic ecosystems.

•••

CHAPTER 4

BIOTIC FACTORS OF DETRITUS DECOMPOSITION

Process of decomposition is controlled by many biotic and abiotic factors (Fig. 4.1). Biotic factors include, substrate quality (chemical composition of litter) and micro- and macroorganisms associated with litter decomposition (Couteaux *et al.*, 1995; Ross *et al.*, 2002) while decomposition (Swift *et al.*, 1979; Silver and Miya, 2001; Gartner and Cardon, 2004).

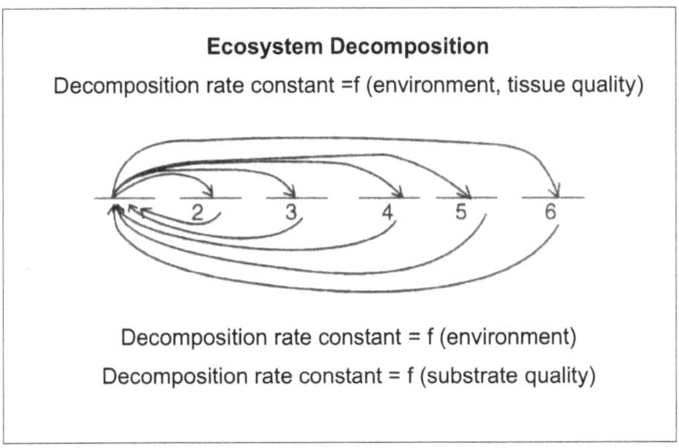

Fig. 4.1. Diagrammatic representation of the role of biotic and abiotic factors in litter decomposition.

Following are the biotic factors that have significant influence on detritus decomposition across ecosystems.

4.1. Substrate quality

It is considered as the major factor affecting decomposition of detritus. Several attributes that constitute substrate quality are described

Litter quality

Differences in detritus decomposition rates of different species have been extensively studied (Aber *et al.*, 1984; Berg *et al.*, 1986; 1991 a, b; Harmon *et al.*, 1986; Corbeels, 2001). Five to ten fold range in decomposition rate of detritus in a given climate is attributed to differences in the substrate quality—measured as the susceptibility of a substrate to decomposition under standardized conditions. For example, animal carcasses decompose more rapidly than plants; leaves decompose more rapidly than wood, deciduous leaves decompose more rapidly than evergreen leaves, and leaves from high nutrient environments decompose more rapidly than leaves from infertile sites (Maclean and Wein, 1978; Cornelissen, 1996; Perez-Harguindeguy *et al.*, 2000). These differences in decomposition rate are due to different types of chemical compounds present in the detritus (Wardle and Lavelle, 1997; Chapin *et al.*, 2002; Wardle *et al.*, 2002) which, for convenience, can be categorized as:

a) labile, metabolic compounds, such as sugars and amino acids,
b) moderately labile structural compounds, such as cellulose and hemicellulose, and,
c) recalcitrant structural material, such as lignin and cutin.

Rapidly decomposing litter generally has higher concentrations of labile substrates and lower concentrations of recalcitrant compounds than the slowly decomposing litter (Table 4.1).

Five interrelated chemical properties (Fig. 4.2) of organic matter determine substrate quality (Schimel, 2001).

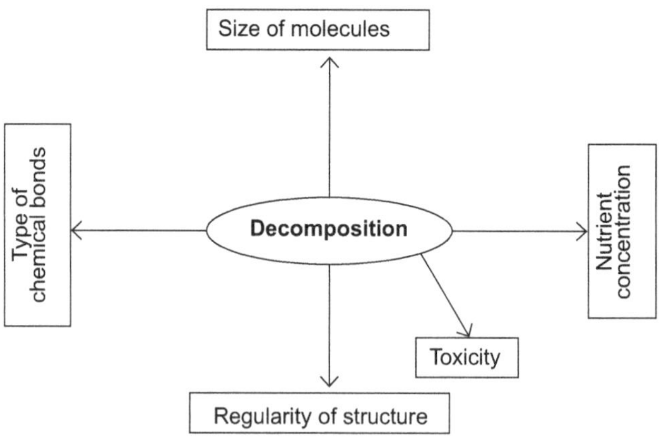

Fig. 4.2. Substrate quality parameters affecting decomposition.

These include: i) size of molecules, ii) types of chemical bonds, iii) regularity of structures, iv) toxicity, and v) nutrient concentrations.

Since the large molecules can not pass through microbial membranes, they must be processed extracellularly by exoenzymes. This limits the degree of control that a given microbe can exert over the detection of substrate availability, delivery of enzymes in response to substrate supply and the efficient utilization of breakdown products. Due to differences in molecular size, sugars and amino acids are metabolized more readily than cellulose and proteins, respectively. Likewise, some chemical bonds are easier to break than others. Ester linkages that bind phosphate to organic skeletons and peptide bonds that link amino acids to form proteins are easier to break than double bonds of aromatic rings. Compounds, like lignin that have a highly irregular structure do not fill the active sites of most enzymes resulting in their slow breakdown compared to compounds like cellulose which consists of chains of regularly repeating glucose units. Some soluble compounds, such as phenolics and alkaloids are toxic and kill or reduce the activity of microbes that absorb them and thereby affect the process of litter decomposition.

Table 4.1. Exponential decay of different chemical constituents comprising plant biomass.

Constituent	% of plant biomass	k (yr^{-1})
Soluble sugars, amino acids, organic acids	10	0.2
Cellulose	15-60	0.08
Hemicellulose	10-30	0.08
Lignin	5-30	0.01

Since organic nitrogen and phosphorus compounds are the major sources of nutrients to support microbial growth, residues, such as straw, that contain low concentrations of these elements may not provide sufficient nutrients to allow microbes to use fully the carbon present in the litter. All these chemical properties influence decomposition, but their relative importance is not well understood. Nonetheless, any of these properties can serve as the predictor of decomposition rate because the properties tend to be strongly correlated with one another. In many studies, N content, or more usually the C:N ratio has been used as an index of substrate quality vis-à-vis decomposition (Melin, 1930; Bocock, 1964; Anderson, 1973b; Flanagan and van Cleve, 1983). Other workers have pointed out that C:P ratio (Heal and French, 1974; Schlesinger and Hasey, 1981; Staaf and Berg, 1982) or lignin content (Fogel and Cromack, 1977; Herman *et al.*, 1977; Meentemeyer, 1978) or lignin: nitrogen content (Taylor *et al.*, 1989; Hobbie, 2000) are better predictors of substrate quality in relation to decomposition. However, the ratio of carbon to nitrogen concentration (C:N ratio) has been frequently used. Litter with low C:N ratio (high nitrogen concentration) generally decomposes quickly (Entiquez *et al.*, 1993; Gholz *et al.*, 2004 But neither the nitrogen concentration of litter nor the nitrogen availability in soil directly influence the decomposition rate in most natural ecosystems (Haynes, 1986; Prescott, 1995; Prescott *et al.*, 1999; Hobbie and Vitousek, 2000). Thus, C:N ratio is not the chemical property that directly controls decomposition in the ecosystems. Furthermore, nitrogen concentration directly limits decomposition rate of organic matter primarily when labile carbon substrates are available to support microbial growth (Haynes, 1986). This is more likely to occur in the rhizosphere than in fresh litter. Under other circumstances, carbon lability rather than nitrogen may be the primary control over decomposition rate (Hobbie, 2000). For instance, in recalcitrant litter the concentration of lignin (or lignin: nitrogen ratio) is often a good predictor of decomposition rate (Berg and Staaf, 1980; Melillo *et al.*, 1982; Taylor *et al.*, 1989) suggesting important role of carbon quality in determining decomposition rates of litter (Chapin *et al.*, 2002). In Sweden, Berg and co-workers (Berg and Staaf, 1980; Berg, 1984, 1986; Berg and Agren, 1984) have independently developed a more refined model of the influence of substrate quality on decomposition. For roots and pine needles, they suggest that nitrogen, and other nutrients, such as phosphorus and sulphur control decay rates during the first phase of decay (<. 30% mass loss), thereafter lignin content becomes progressively more important.

Supposition of Berg (1986) that N control of litter decay rate should be succeeded by lignin control is supported by Taylor *et al.* (1989) who indicated that this relationship extends to a broad range of leaf litter types. During litter decomposition, the relative lignin concentration apparently increases not only owing to its low degradation rate (Berg *et al.*, 1987) but also because the conventional lignin analytical methods do not distinguish between true lignin and partially humified products. However, the rate retarding influence of lignin on litter decomposition appears to depend on climate (Couteaux *et al.*, 1995). It was found that in late decomposition stages the effect of higher lignin concentrations was less in harsh climate (e.g., at the arctic circle) compared to warmer and wetter regions where higher lignin concentrations have predominant effect (Couteaux *et al.*, 1995). Also nitrogen availability during decomposition of litter has significant effect. Studies of Hobbie (2000) in Hawaiian montane forests revealed that N deposition promotes faster litter decomposition, more rapid N-cycling through litter and faster transient C loss in ecosystems characterized by low lignin litter. Nitrogen deposition in ecosystems characterized by high lignin litter has little effect on litter decomposition. Effect of litter quality on decomposition rate often depends on the age of the litter. For instance, quick loss of labile carbon from high quality litter leaves behind litter that may have lower decomposition potential (Berg and Ekbohm, 1991).

Availability of below-ground resources, like nutrients and water is another major ecological control over litter quality. Rapidly growing plants from high resource sites typically produce litter that decomposes quickly because the same morphological and chemical traits that promote net primary productivity also regulate decomposition (Hobbie, 1992). Both net primary productivity and decomposition are enhanced by high allocation to leaves and production of leaves with a short life span. These tissues decompose rapidly because they have high concentrations of labile compounds, such as proteins and low concentrations of recalcitrant cell wall components, such as lignin (Reich *et al.*, 1997). Consequently, species from productive sites produce litter that decomposes rapidly (Cornelissen, 1996). However, species differences in litter quality make up an important mechanism by which plant species affect ecosystem processes (Hobbie, 1992) and are excellent predictors of landscape patterns of litter decomposition (Flanagan and van Cleve, 1983).

Quality of the roots (root chemistry) is also regarded as the dominant factor controlling patterns in decomposition rates at a global scale. Effect of initial chemistry of roots of different diameter classes vis-a-vis decomposition is easily discernible in Table 4.2.

Table 4.2. Initial litter quality by root diameter class and root decomposition. (Values are means± SE).

Variable	<2 mm	2-5 mm	>5 mm
k-values	0.83± 0.15	0.52±0.09	0.18±0.03
C:N	67± 6	104±25	156±19
Lignin:N	37± 3	53±5	73±11
Lignin:P	565± 48	380±48	720±157
Ca (mg/g)	2.32± 0.43	2.91±0.25	0.87±0.22
N (%)	0.94±0.05	0.59±0.05	0.45±0.06
P (%)	0.05±0.003	0.07±0.004	0.05±0.003
Lignin (%)	23.6±0.8	22.4±0.8	22.7±1.0

Source: Silver and Miya (2001)

Among the litter quality indices, root calcium is best predictor of decomposition rates of fine roots (Table 4.3). Several explanations, such as the involvement of root Ca in determination of soil conditions, influence on microbial activity and its regulatory role in many physiological processes have been put forth to explain its role in root decomposition.

C:N ratio (Table 4.3) is also a good predictor of root decomposition. Theoretically, the optimum C:N ratio for microbial growth is approximately 25, but fungi and bacteria can decompose substrates with much higher ratios as well. Substrates with C:N ratios of <20 decompose rapidly with release of NH_4 while as materials with C:N ratio of 25-75 though decompose quickly but NH_4 is not released due to its immobilization by microbes and complexation of proteins by polyphenols. Substrates with'high C:N ratio (>75) are often resistant to decomposition due to higher amounts of lignin, condensed tannins and terpenes, as well as less amount of available N for decomposer organisms.

Plant life form through its influence on litter chemistry also influences root decay rates (Table 4.4). Rate of decomposition is in the order of graminoids > broadleaved species > coniferous species.

Table 4.3. Relationship between substrate quality parameters and decomposition (k).

Parameter	Regression equation	R^2	Significance
Fine roots			
C:N	ln(k)=3.16-0.90*ln(C:N)	0.21	P<0.05
Lignin:N	ln(k)=1.47-0.66*ln(lignin:N)	0.20	P<0.001
Lignin:P	ln(k)=2.99-0.58*ln(lignin:P)	0.10	P<0.05
Ca mg/g	ln(k)=-1.22+0.70*ln(Ca)	0.55	P<0.001
Lignin (%)	ln(k)=4.61-1.72*ln(lignin)	0.16	P<0.05
All diameters			
C:N	ln(k)=3.92-1.12*ln(C:N)	0.34	P<0.001
Lignin:N	ln(k)=1.27-0.60*ln(lignin:N)	0.20	P<0.001
Lignin:P	ln(k)=2.36-0.53*ln(lignin:P)	0.12	P<0.001
Ca mg/g	ln(k)=-1.33+0.49*ln(Ca)	0.31	P<0.001
P(%)	ln(k)=0.19+0.37*ln(P)	0.03	P<0.05
Lignin (%)	ln(k)=3.47-1.40*ln(lignin)	0.18	P<0.001

Source: Silver and Miya (2001)

Table 4.4. Root decomposition in relation to different plant life forms and their initial substrate quality. (Values are means± SE).

Variable	Graminoid	Broadleaf	Conifer
Fine roots			
k-values	1.48=0.33	0.46±0.06	0.17=0.02
C:N	70±10	55=4	93=20
Lignin:N	26±3	44±5	41=9
Lignin:P	439=36	646=82	628=161
Ca (mg/g)	4.88=1.43	2.10=0.42	0.95±0.24
N (%)	0.98±0.10	0.93=0.07	0.82=0.10
P(%)	0.05=0.005	0.06=0.004	0.05=0.008
Lignin (%)	17.9±0.9	27.6=0.7	2601±1.2
All size classes			
k-values	1.41±0.27	0.44±0.04	0.30±0.02
C:N	66±8	92±14	132±17

Lignin:N	26±3	44±4	67±4
Lignin:P	414±56	698±80	431±36
Ca (mg/g)	3.79±0.89	1.95±0.37	2.34±0.22
N (%)	0.90±0.08	1.95±0.37	0.47±0.04
P(%)	0.06±0.005	0.84±0.05	0.06±0.003
Lignin (%)	18.2±1.0	26.8±0.9	24.0±0.5

Source: Silver and Miya (2001)

Soil organic matter (SOM) quality

Age as well as the initial quality of SOM influence its rate of decomposition. Through fragmentation by soil invertebrates and chemical alterations brought about by microbes, litter is converted to SOM. With the passage of time SOM ages and several changes occur in it that alter its rate of decomposition. C: N ratio, for example, declines as decomposition proceeds because carbon is respired and some of the mineralised nitrogen is incorporated into humus. Decline in C:N ratio is not, however, an indicator of increased nitrogen availability because nitrogen becomes incorporated into aromatic rings and other chemical structures that are recalcitrant. As in litter, carbon quality of the SOM is a better predictor of decomposition rate than is C:N ratio or the nitrogen concentration of SOM (Berg and Staaf, 1980; Melillo *et al.*, 1982).

Site differences in nutrient availability also influence SOM decomposition primarily through their effects on the carbon quality of litter and SOM, rather than through direct nutrient effects on SOM decomposition. Sites with high productivity and litter quality typically produce a low lignin SOM that decomposes readily (van Cleve *et al.*, 1983). As in the case of fresh litter, SOM decomposition rate does not show a consistent response to nutrient addition (Haynes, 1986; Fog, 1988) suggesting that nutrients seldom directly regulate SOM decompoiition. Increase in decomposition of SOM in response to nitrogen addition occurs when the organic matter consists of labile carbon substrates. Ploughing of straw into agricultural soils (Mary *et al.*, 1996) or enhanced root exudation by elevated CO_2 (Hu *et al*, 2001) exemplify such a pattern in SOM decomposition.

The heterogeneous nature of SOM makes it difficult to identify the chemical controls over its decomposition. It is a mixture of organic compounds of different ages and chemical composition. Components of SOM include fragments of recently shed root and leaf litter, together with soil organic matter that is thousands of years old (Oades, 1989). These different aged components of SOM can be separated by density centrifugation, as recently produced particles are less dense than older ones and are less likely to be bound to mineral particles. Soils in which a large proportion of the SOM is in the light fraction generally have higher decomposition rates (Robertson and Paul, 2000). SOM as a whole typically has a residence time of 20 to 50 years, although this can range from 1 to 2 years in cultivated fields to thousands of years in environments with slow decomposition rates. Even in a single soil, different chemical fractions of SOM have residence times ranging from days to thousands of years (Chapin *et al.*, 2002). Computer simulations of decomposition rate capture ecosystem carbon dynamics more effectively when they distinguish among the different soil carbon pools (Parton *et al.*, 1993; Clein *et al.*, 2000).

4.2. Decomposer community

Decomposer community is predominantly composed of microorganisms (bacteria and fungi) and invertebrate animals (Boer *et al.*, 2005).

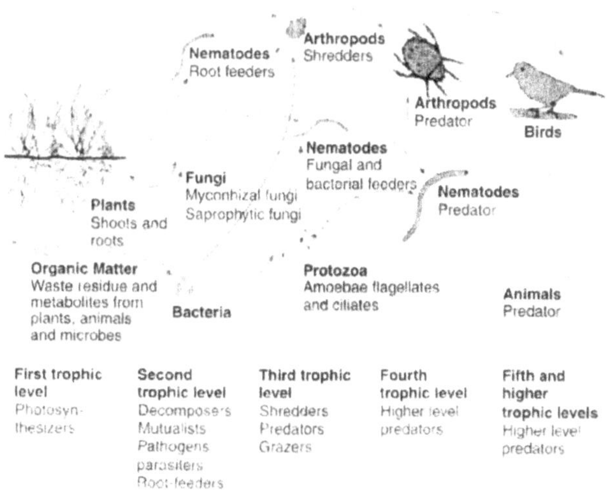

Fig. 4.3. Soil food web — a diagrammatic representation.

Composition of decomposer community influences decomposition rates (Agren *et al.*, 1996) with soil animals, in particular, accelerating the decomposition rate either by modifying the physico-chemical environment through fragmentation and transformation of litter (Chapin *et al.*, 2002) or indirectly by stimulating microbial activity through grazing of bacteria and fungi (Anderson and Ineson, 1983; 1984). Diagrammatic representation of the soil food web is given in Fig. 4.3.

Soil animals

The soil animals are usually classified on the basis of body width (Table 4.5) as microfauna, mesofauna and macrofauna.

Table 4.5. Size classification of organisms in decomposer food webs by body width.

Group	Classes	Body width
Microflora and microfauna	Bacteria Fungi Nematoda Protozoa Rotifera	Up to 100 pm
Mesofauna	Acari Collembola Protura Diplura Symphyla Enchytraeidae Chelonethi Isoptera	100 μm-2 mm
Marcro- and megafauna	Opiliones Isopoda Amphipoda Chilopoda Diplopoda Megadrili Coleoptera Araneida Mollusca	2 mm-20 mm

Source: Swift *et al.* (1979)

Microfauna include the smallest animals (size less than 0.1 mm), such as nematodes, protozoans viz., ciliates, amoebae and some mites (Wallwork, 1976; Lousier and Bamforth, 1990). Since they move through water films on the surface of soil particles, protozoans and nematodes are more sensitive to water stress than other species. Consequently their populations fluctuate dramatically both spatially and temporally due to drying and wetting events and predation (Beare *et al.*, 1992). The single celled protozoans are usually mobile and are voracious predators of bacteria and other microfaunal species (Lavelle *et al.*, 1997) and ingest their prey primarily by phagocytosis. Protozoans are particularly important predators in the rhizosphere and other soil microsites with rapid bacterial growth rates (Coleman, 1994). This preferential grazing by protozoa on bacteria (even on particular species of bacteria), alters microbial community composition causing reduced bacterial: fungal ratio compared to soils from which protozoa are excluded. It is assumed that this type of grazing preference maintains the bacterial population in a youthful state which aids in maintaining decomposition activity (Griffiths, 1994). Based on the evidence of increased respiration, ammonification and siderophore production, amoebae are known to have stimulatory effect on bacterial cultures (Griffiths, 1994).

Nematodes, an abundant and trophically diverse group, include species that specialize on bacteria, fungi, roots or other soil animals. Bacterial feeding nematodes in forest litter can consume about 80 g m^{-2} yr^{-1} of bacteria, resulting in mineralization of 2 to 13 g m^{-2} yr^{-1} of nitrogen representing a substantial proportion of the nitrogen that cycles annually through the soil (Anderson *et al.*, 1981).

Mesofauna (0.1 to 2 mm in length), a taxonomically diverse group, includes the animals that have the greatest effect on detritus decomposition. They fragment and ingest litter coated with microbial biomass, producing large amounts of faecal material with a greater surface area and moisture holding capacity than original litter (Lavelle *et al.*, 1997). This altered litter environment is more favourable for decomposition. Mesofauna selectively feed on litter that has been conditioned by microbial activity. Within mesofauna, Collembolla, the small insects, feed primarily on fungi, whereas mites (Acari) are a more trophically diverse group of spider like animals that consume decomposing litter or feed on bacteria and/or fungi. So microbial grazing by mesofauna regulates decomposition of organic matter by altering community composition (Wasilewska *et al.*, 1975; Trofymow and Coleman, 1982; Whitford *et al.*, 1982 a; Yeates ftnd Coleman, 1982; Seastedt, 1984).

Macrofauna are largest in size (greater than 2 mm) but least abundant per unit area (Neher, 1999). In contrast to micro- and mesofauna which do not have the ability to reshape soil and inhabit soil pores of 25-100 mm diameter (Griffiths, 1994), macrofauna, such as earthworms and termites are termed as ecosystem engineers because they alter resource availability by modifying the physical properties of soil and litter (Jones *et al.*, 1994). Some of them fragment litter like the mesofauna (Lavelle *et al.*, 1997), others burrow or ingest soil, reduce soil bulk density, break up soil aggregates, and increase soil aeration and the infiltration of water (Beare *et al.*, 1992). Earthworms create patterns of soil structure that promote or constrain the activities of soil microbes and other soil animals (Chapin *et al.*, 2002). In temperate pastures, earthworms process 4 kg M^{-2} yr^{-1} of soil, and move 3 to 4 mm of new soil to the ground surface each year (Paul and Clark, 1996). During this processing, digestive tract secretion of earthworms stimulate microbial activity but many of the soil organisms are lysed and digested during their passage through the gut (Chapin *et al.*, 2002). Cortez and Bouche (1998) noted that earthworms make litter more palatable by their behaviour. During the initial stages of decomposition, the litter is ploughed by earthworm casts involving both increase in the microbial activity and preliminary microbial litter decomposition which ulitmately results in enhanced litter decomposition (Wright, 1972; Cortez and Hameed, 1988).

Notwithstanding the nature and role of soil animals as elucidated above, distinct physiological and environmental requirements drive soil organisms to play widely different roles in soil ecosystem processes, such as promotion of plant growth, receiving, holding and releasing water, recycling carbohydrates and nutrients through mineralization, acting as environmental buffer and transferring energy in food chain (Neher, 1999). Smith and Bradford (2003) studied dependence of litter decomposition on soil fauna and hypothesized that higher quality litter would decompose more rapidly, and that this effect would occur earlier when larger fauna were present because they are mobile than smaller organisms and so potentially can colonize litter more rapidly. Loss or exclusion of soil invertebrates can reduce decomposition rate and, therefore, nutrient cycling substantially, indicating the important role of soil animals in decomposition process (Swift *et al.*, 1979; Verhoef and Brussaard, 1990). Since, soil animals account for only about 5% of soil respiration, their major effect on decomposition is by the enhancement of microbial activity through litter fragmentation (Wall *et al.*, 2001) rather than their own processing of energy derived from detritus.

Other fauna like micro-arthropods and Enchytraeidae have indirect effects on decomposition. Such organisms fragment detritus and increase surface area for further microbial attack (Berg and Pawluk, 1984; van Vliet *et al.*, 1995). Examples of invertebrate-microbe mutualisms are common in soils. For instance, most soil invertebrates (e.g., termites and earthworms) do not seem to possess suitable enzymes to directly digest resources, such as cellulose, lignin, tannin and humic complexes in soil; instead the enzymes seem to be produced by ingested microbes rather than by invertebrates themselves (Lavelle *et al.*, 1995). In addition to soil organisms, food web structure also influences decomposition rate (Setala and Huhta, 1991; Bengtsson *et al.*, 1995) but according to Andren *et al.* (1995) the ecosystem function depends more on the presence and abundance of a particular functional group rather than the complexity of food webs or biodiversity alone. The "snow chain" hypothesis predicts a succession of decomposer organisms that respond to changes in substrate quality (Neher, 1999). Once the substrate is decomposed somewhat and changed qualitatively, number of micro-arthropods, nematodes and predators increase with a subsequent decline in number of bacterial feeding nematodes and enchytraeids which are initially present in large quantities (Andren et *al.*, 1995). Thus, it is evident that succession occurs among soil animals during transfer of energy in the food chain (Neher, 1999). Slower decomposition rates have been reported for systems without mites (treated with the insecticide chlordane) than with mites (no insecticide applied) (Neher, 1999) and in the former nematode grazing decreased the decomposition rate. Presence of mites reduce nematode populations and release microbial populations from predation, which results in faster decomposition rate. Consequently, predatory mites maintain nematode grazer population at levels that presumably increase microbial activity and decomposition rates. Studies of Beare *et al.* (1992) and Bradford *et al.* (2002) showed that rates of litter decomposition and nitrogen dynamics in ecosystem can be tied to differences in composition of decomposer communities and their trophic interactions.

Soil microorganisms

The decomposer microflora consists primarily of bacteria, fungi, actinomycetes and yeasts (Berg *et al.*, 2001). These microorganisms are very important in bringing about decomposition. Macfadyen (1963) has shown that 80% of the energy flows through the terrestrial decomposer community and Reichle *et al.* (1975) also observed that 99% of the energy and caloric input to the forest passes through microflora. Thus, it is the single most important group in the annual turnover of energy captured in photosynthesis (Singh and Gupta, 1977).

During the process of decomposition there occurs a succession in the nature of organisms and this succession of microorganisms in litter decomposition is primarily determined and controlled by the abilities of the microorganisms present to utilize available organic compounds, and partly also by the response of the organisms to change in the environment caused by the decomposition (Alexander, 1964).

The role played by different groups of microorganisms in decomposition is summarized below.

Terrestrial fungi

Fungi are next only to bacteria in abundance in most soils (Subba Rao, 2000) and are the main initial decomposers of terrestrial dead plant material and together with bacteria, account for 80-90% of the total decomposer biomass and respiration (Chapin *et al.*, 2002). The number of fungal species is at least 100,000 or as many as 250,000 (Pugh, 1974). Of these, a minority in normal circumstances appear to require the presence of a living host. These include various mycorrhizal fungi and the obligate biotrophs which parasitize their hosts without killing them while the remaining fungi are either unspecialised necrotrophs (facultative parasites able to live on the dead remains of the host) or saprophytes. The latter group plays a significant role in the decomposition of plant litter. Fungi have networks of hyphal filaments that enable them to grow into new substrates and transport materials through the soil over distances of centimetres to meters. These hyphal networks enable fungi to acquire their carbon in one place and their nitrogen in another, much the same way as plants gain CO_2 from the air and water and nutrients from the soil.

Fungi, by virtue of their diverse enzyme machinery, are capable of degrading a wide variety of substrates (Rabinovitch *et al.*, 2004) and hence play a significant role in litter decomposition which has been described in detail by Harley (1971). Undoubtedly, fungi are the principal decomposers of fresh plant litter because they secrete enzymes that enable them to penetrate the cuticle of dead leaves or the suberised exterior of roots to gain access to the interior of a dead plant organ and proliferate within and between dead plant cells. At a smaller scale, some fungi gain access to the nitrogen and other labile constituents of dead cells by breaking down the lignin present in cell walls. This energy investment in lignin degrading enzymes serves primarily to gain access to the relatively labile contents of the interior of cells. Fungi produce hyphae with a dense concentration of cytoplasm when there is adequate substrate to support growth. The hyphae contain more vacuoles (and proportionally less cytoplasm) when resources are scarce. This flexible growth strategy enables fungi to grow into new areas to explore for substrate, even when current substrates are exhausted. Substantial proportions (perhaps 25%) of carbon and nitrogen used to support fungal growth are transported from elsewhere in the hyphal network, rather than being absorbed from the immediate environment where the fungal growth occurs (Mary *et al.*, 1996).

Fungi enjoy a competitive advantage over bacteria in decomposing tissues with low nutrient concentrations because of their ability to import nitrogen and phosphorus from distant substrates. In particular, the white rot fungi specialize in lignin degradation in logs, where as brown rot fungi cleave some of the side chains of lignin but leave the phenol units behind. White-rot fungi are generally out competed by more rapidly growing microbes when nitrogen is abundant. Thus, nitrogen additions have little effect (or sometimes a negative effect) on white-rot fungal decomposition of wood. Fungi account for 60%-90% of the microbial biomass in forest soils, where litter frequently has a high lignin and low nitrogen concentration (Chapin *et al.*, 2002).

Mycorrhizal fungi, though get most of their carbon from host plant roots, can also play a role in decomposition by breaking down proteins into amino acids that support fungal growth and are transferred to their host plants as well (Read, 1991; Koide *et al.*, 2005). Mycorrhizal fungi also produce cellulases to gain entry into plant roots but it is uncertain whether these cellulases participate in decomposition of dead organic matter. Indeed, some studies suggest that ectomycorrhizal fungi or ectomycorrhizal roots promote decomposition or mineralization (Zhu and Ehrenfeld, 1996). However, Gadgil and Gadgil (1971; 1975) suggested that ectomycorrhizal fungi might actually retard decomposition by competing with saprophytic microorganisms and such antagonistic interaction is known as 'Gadgil' effect. If two-sided competition between ectomycorrhizal fungi and saprotrophic microorganisms occurs, which is at least theoretically possible, then ectomycorrhizal fungi may retard decomposition by limiting the activities of the saprotrophs.

The range of substrates upon which fungi can grow is dependant on their enzyme systems. While virtually all fungi can utilize glucose, increasing complexity of the substrate requires the production of particular enzymes, the demonstration of which, in the laboratory at least, indicates the potential to use that substrate in the field. As cellulose forms such a large part of plant remains, cellulases are very important in nature. The occurrence of cellulose decomposing fungi has been studied by many workers (Siu, 1951; Kelman, 1967). Other enzyme systems that are important in plant litter decomposition include cutinase and pectinase, found in plant pathogens and phylloplane fungi, amylase, lipase, proteinases and saccharase which degrade plant cell contents. The decomposition of lignin involves phenol oxidase and phenolases produced by white rot fungi which are involved in the breakdown of lignin and other polyphenols (Hurst and Burges, 1967). Polyphenols and tannins in particular are responsible for the unplatability of many leaves to the soil fauna. They are broken down to catechols and gallic acid, for example by *Penicillium expansum* (Pugh, 1974).

Actinomycetes

Actinomycetes share characters of both bacteria and fungi and are commonly known as "ray fungi" because of their close affinity with fungi (Dubey and Maheshwari, 1999). They are branched gram-positive bacteria with aerial and specialized hyphae. These hyphae although analogous to those of fungi, are considerably narrower (generally 0.5-1.2 m diameter) and are composed of prokaryotic cells (Nester *et al.*, 1995). Populations of soil actinomycetes are greater in grasslands and pastures than in cultivated land. In temperate zones the number of actinomycetes ranges from 10^{5} to 10^{8} per gram of dry soil (DubeyandMaheshwari, 1999). Themostlimitingfactorwhichgoverns their abundance in soil is pH. Their luxuriant growth is favoured by neutral or alkaline pH (6.0 to 8.0) (Garrett, 1981).

Role of actinomycetes in litter decomposition has received far little attention compared to bacteria and fungi. The meagre evidence available does suggest that the actinomycetes play a limited yet important role in the cycling of organic materials and minerals (Goodfellow and Cross, 1974). Most of the actinomyetes isolated from litter have been identified as streptomycetes (Goodfellow and Cross, 1974). Jensen (1971) estimated that the number of streptomycetes account for less than 2% of the total bacterial flora of *Fagus* litter, a result in agreement with earlier findings of Wolniewicz-Czerwinska (1956). Low counts of actinomycetes have also been observed in *Juncus* litter (Latter *et al.*, 1967), highly acid peat soils (Ishizawa and Araragi, 1970) and in acid, as opposed to alkaline and neutral soils (Corke and Chase, 1964; Davies and Williams, 1970). These findings suggest that acidic conditions represent one of the prime factors limiting the presence of actinomycetes (Brock, 1969). There are some data which suggest that actinomycetes other than streptomycetes are common in litter. Numerous strains of *Streptosporangium* have been reported in *Qurecus* and Fagus forest litter (van Brummelen and Went, 1957). Likewise, Ruddick and Williams (1972) found that strains of *Stmgriseus* were common on the bodies of the indigenous litter fauna, and formed up to 20% of streptomycetes population in Pinus litter. The strains of *Oerskoviae* and *Promicromonosporae* have been isolated from decaying organic matter (Lechevalier, 1972). Although mycobacteria have not been specially isolated from litter, they do seem to be part of the normal bacterial flora of soil (Beerwerth and Schurmann, 1969).

The ability to grow at temperatures above 40°C is found in several genera of actinomycetes, especially in Thermoactinomyces, *Thermonzonospora, Streptomyces, Pseudonocordia* and *Micropoly-spora*.Many strains in these taxa possess amylolytic and celluloytic properties (Henssen, 1957 a, b; Fergus, 1969; Kuo and Hartmann, 1966; Stutzenberger *et al.*, 1970; Stutzenberger, 1971). It seems that thermophilic actinomycetes can grow in litter and surface layers of soil in temperate regions as long as the sites are exposed to sunlight (Åpinis, 1965).

It is generally accepted that actinomycetes can utilize a wide range of residues of plants and animals. Streptomycetes and other allied actinomycetes grow on roots, twigs, leaf debris and derive some of their nutrients from plant remains. It is also on record that some streptomycetes derive their nutrients from living or dead hyphae of the primary fungal colonizers. The lysis of fungal hyphae by streptomycetes has also been reported (Skuj ins *et al.*, 1965; Jones *et al.*, 1968; Howard and Gupta, 1971) and these organisms exist for long periods in soil and litter as dormant spores, as their vegetative growth phase is superseded by spore phase (Williams *et al.*, 1971).

The common actinomycetes, such a Streptomycetes, Nocardias and Micromonosporae, generally utilize cellulose (Waksman, 1919; Fergus, 1969; Ishizawa and Araragi, 1970), chitin (Veldkamp, 1955; Jeuniaux, 1955; Okafor, 1966; Gray and Baxby, 1968), keratin (Noval and Nickerson, 1959; Kuchaeva *et al.*, 1963; Goodfellow, 1971), lectin (Bilimoria and Bhat, 1961; Knosel, 1970), xylan (Kusakabe *et al.*, 1969) and oxalic acid (Jagnow, 1957). In addition, Nocardias and strains of the Rhodochorus complex can degrade a large range of carbon compounds including long chain fatty acids and hydrocarbons (Raymond *et al.*, 1967; Goodfellow, 1971). While Nocardias have been implicated in decomposition of humic acids, streptomycetes are known to contribute to the formation of humus (Kuster, 1967; Kutzner, 1968) though little is known of this complex process.

In conclusion it may be stated that the actinomycetes are probably an ephemeral vegetative component of the soil and litter microflora and, like other groups of soil bacteria, show rapid growth when suitable food material such as plant litter is present. They have a lower competitive ability than the fungi and rapidly growing bacteria but can germinate and grow on fresh plant material. However, actinomycetes can survive as spores and have a marginal advantage over some of the associated fungi, for the spores can later germinate and exhibit limited growth when adjacent to autolysing fungal hyphae or more recalcitrant plant debris (Goodfellow and Cross, 1974). Thus, actinomycetes play a minor role in decomposition of the total litter added to soil (Gray and Williams, 1971) but form an integral part of a balanced biological community.

Bacteria

It is the most abundant unicellular prokaryote group. The number varies between 10^8 to 10^{10} cells per gram of dry soil (Dubey and Maheshwari, 1999). The soil bacteria are usually divided into two functional groups:

a) the indigenous organisms, whose number in soil is supposed to remain unaffected by the litter, and,
b) the zymogenous organisms, which are actively involved in litter decomposition and which, therefore, increase to very high numbers during periods of litter decomposition but do not disappear completely from the soil when decomposition slows down. The leaching of water soluble compounds from recently fallen litter causes an activation of retained zymogenous organisms, which take over the next step of degradation, especially in case of easily decomposable polymeric compounds, including pectin, hemicelluloses, celluloses and chitin (Gyllenberg and Eklund, 1974).

Though the abilities to degrade the polymeric compounds are widespread among fungi and actinomycetes but among bacteria these activities are restricted to few genera or only a few species in certain genera (Gyllenberg and Eklund, 1974). In addition, the prevalent environmental conditions may limit bacterial activity. Consequently, in most environments, fungi constitute the primary decomposer population of plant material, whereas bacteria appear to have secondary role. This particularly concerns forest soils where the predominance of fungi as decomposing agents is partly due to high acidity and partly because of the inhibition of bacterial enzymes by tannins present in the litter (Henis *et al.*, 1964; Basaraba and Starkey, 1966). Bacteria in turn exhibit considerable activity in the breakdown of fungal mycelium and other similar materials (Gyllenberg and Eklund, 1974).

Notwithstanding their secondary importance, a wide range of bacterial types are present in soils. The bacterial communities that coat soil aggregates exhibit a surprisingly complex structure and are often present as biofilms—a microbial community embedded in a matrix of polysaccharides secreted by bacteria (Chapin *et al.*, 2002). The bacteria in biofilms often act as a consortium i.e., a group of genetically unrelated bacteria, each of which produces only some of the enzymes required for breakdown of complex macromolecules. The breakdown of these macromolecules to the point that soluble products are released requires the coordinated production of exoenzymes by several types of bacteria (Chapin *et al.*, 2002).

Most reports concerning role of bacteria in litter decomposition are descriptive and information is given only of relative numbers of various bacteria in decaying organic matter (Gyllenberg and Eklund, 1974). For a few groups of bacteria more exact information is available. Among these, is a group of fluorescent pseudomonads, which are distributed unevenly in soil and are associated particularly with organic matter (Rovira and Sands, 1971). The bacilli is another group of bacteria associated with the decay of plant residues, especially Bacillus circulans which occurs frequently in acid forest soils (Holding *et al.*, 1965; Goodfellow *et al.*, 1968). Pseudomonads can be considered as particularly important in the decomposition of the aromatic compounds released during different stages of litter degradation (Dagley, 1967). Other bacteria which have been reported as decomposers of aromatic compounds are *Agrobacterium, Arthobacter, Achromobacter* and *Cellulomonas* (Stevenson, 1967; Kunc, 1971). Some proteolytic bacteria are also present in soils which are capable of utilizing polypeptide chains bound to the aromatic skeleton (Haider *et al.*, 1965). Thus, the role of bacteria in plant litter decomposition is partly a direct breakdown of litter constituents, and partly indirect degradation of the organic material which accumulates as a result of litter decomposition (Gyllenberg and Eklund, 1974).

•••

CHAPTER 5

SOIL ENZYMES AND DETRITUS DECOMPOSITION

5.1. Nature and role of soil enzymes

General account

All biochemical activity during the process of decomposition is dependent upon or related to the presence of enzymes (Stevenson, 1994). In fact, many biological transformations occurring in soils are catalyzed by enzymes found outside living organisms (Sinsabaugh *et al.*, 2002). Such enzymes are often named as free enzymes (Kiss *et al.*, 1972), soil enzymes (Kuprevioh and Shcherbakova, 1966) or sometimes extracellular enzymes (Dickson and Pugh, 1974). The two main types of soil enzymes are:

a) constitutive which are usually produced by microorganisms and are always present in soil, such as dehydrogenases and urease (Dar, 1995; Subba Rao, 2000),
b) inducible which are produced by microbes in presence of specific chemical compounds. Examples include cellulase and phosphatase (Dar, 1995; Subba Rao, 2000).

Soil enzymes play an essential role in catalysing reactions necessary for organic matter decomposition and nutrient cycling (Senthil Kumar *et al.*, 1997; Ajwa *et al.*, 1999; Kourtev *et al.*, 2002), energy transfer, environmental quality and crop productivity (Dick, 1994; Tabatabai, 1994). Their activities are generally considered to be a direct expression of soil biological activity (Ladd, 1985) as many studies (Tabatabai, 1977; Speir *et al.*, 1980; Frankenberger and Dick, 1983; Perucci *et al.*, 1984; Tate, 1987; Dick *et al.*, 1988a, b) have reported significant correlation between soil enzyme activities and microbial biomass. Hence, soil enzyme activities are often used as indices of microbial activity (Dick and Tabatabai, 1992), nutrient cycling and microbial populations (Ladd, 1985; Kumar *et al.*, 1992; Dick, 1994).

Chapin *et al.* (2002) pointed out that most soil microbes produce enzymes (proteases and peptidases) that breakdown proteins into amino acids. These breakdown products are readily absorbed by microbes and used either to produce microbial protein or to provide respiratory energy. As proteases are attacked by other proteases, their lifetime in the soil is short and soil protease activity tends to mirror microbial activity (Chapin *et al.*, 2002). Phosphatases, which cleave phosphate from organic phosphate compounds, are, however, more long lived and their activity in soils is correlated more strongly with availability of organic phosphate than microbial activity (Kroehler and Linlcins, 1991). Cellulose, the most abundant chemical constituent of plant litter (Chapin *et al.*, 2002), requires three separate enzyme systems for its breakdown (Paul and Clark, 1996): endocellulases that breakdown the internal bonds to disrupt the crystalline structure of cellulose; exocellulases which then cleave off disaccharide units from the ends of chains forming cellobiose which is then absorbed by microbes and broken down intracellularly to glucose by cellobiase. Some soil microbes, including most fungi can produce the entire suite of cellulase enzymes (Chapin *et al.*, 2002). However, some bacteria, do not produce all the cellulose degrading enzymes and hence function as part of the microbial consortia to gain energy from cellulose breakdown. Lignin is degraded slowly because only some organisms (primarily fungi) produce the necessary enzymes, and such enzymes are produced only when other more labile substrates are unavailable.

Currently decomposition is considered as a successional loop wherein substrate selects the microbial community, which produces extracellular enzymes that degrade and modify the substrate, which in

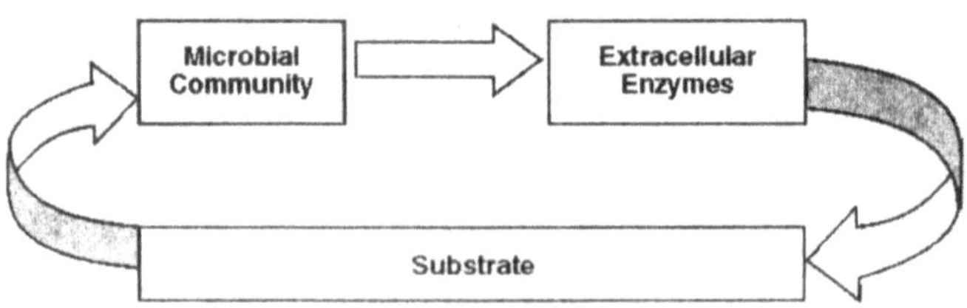

Fig. 5.1. The decomposition process as a successional loop.

turn, drives the succession of microbial community during detritus decomposition. Thus, extracellular enzymes provide a link between substrate composition and microbial community metabolism (Fig. 5.1).

Detritus decomposition, as stated earlier (Chapter 3), has been widely modelled as a process that follows first order decomposition kinetics (Molina *et al.*, 1990; Chertov and Komarov, 1997; Li, 1996) but recently a simple theoretical model has been developed by Schimel and Weintraub (2003) wherein the behaviour of decomposition-microbial growth system is changed from first order kinetics to exoenzyme catalysed decomposition (see Chapter 3 for details).

In fact, Parnas (1975) was among the first to present a model of litter decay that was controlled by differential acquisition of macronutrients by decomposers. Sinsabaugh and Moorhead (1994) extended this approach by developing an explicit model of microbial allocation of resources among community indicator enzymes **(MARCIE)**, with estimation of timing and levels of activity for particular enzymes based on energy and nutrient availabilities. Sinsabaugh *et al.* (1991) have shown that temporally integrated rates of enzymatic activities correlate with mass loss pattern in litter and Sinsabaugh and Moorhead (1994) used the MARCIE model to simulate overall patterns of litter decay.

Generally the relationship between soil enzymatic activity and microbial growth and activity is complicated by the existence of stabilized extracellular enzymes and inclusion of enzymes contained within inactive or resting cells in the total enzymatic activity measurements (Tate, 1987). Under most conditions, enzymatic activity directly associated with intact microbial cells, and not expressed outside of the living cell, can be anticipated to relate in part with overall microbial respiration (Casida, 1977; Skujins, 1973). Recent studies (Sinsabaugh *et al.*, 1991; Sinsabaugh and Moorhead, 1994) have shown that patterns of litter decay are correlated with activities of key classes of enzymes and as such decomposition can be modelled as a function of the activity of these enzymes. Since the substrate quality varies with time, hence the activity of different enzymes also varies over time (Fig. 5.2) though other biotic and abiotic factors might also be responsible for such temporal variation in enzyme activity. In view of the above stated importance of exoenzymes vis-a-vis litter decomposition several studies have focussed attention on this aspect both in terrestrial and aquatic ecosystems.

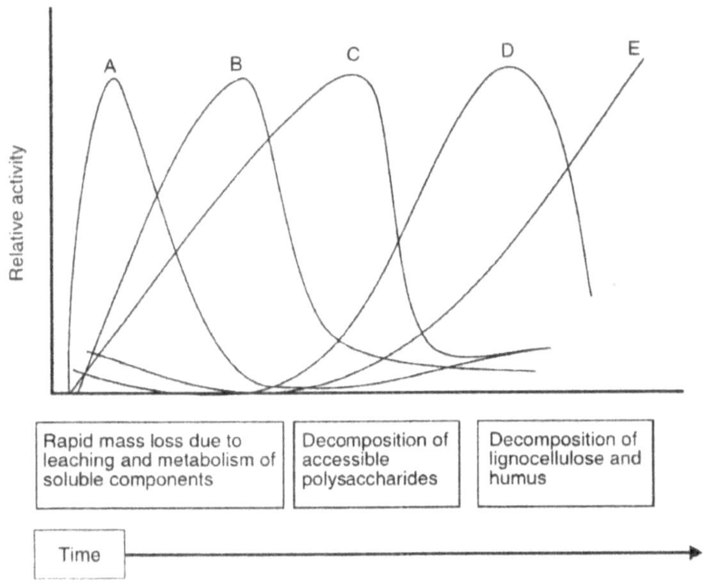

Fig.5.2. Changes in the activity of different enzymes during decomposition of herbaceous plant litter. (A) invertase, α-glucosidase; (B) β-1,4-exoglucanase (exocellulase), (C) β-1,4-endoglucanase (endocellulase), (D) (poly) phenol oxidase, (E) peroxidase.

Role in terrestrial ecosystems

In terrestrial ecosystems most of the studies have been carried out primarily to relate activity of enzymes to litter composition or microbial dynamics at a fine scale. Some workers, however, have followed large-scale comparative ecosystem approach to investigate the role of microbial enzymes in litter decomposition. While studying decomposition of senescent leaves of *Cornus florida, Quercus prinus*, and *Acer rubrum* in a deciduous woodland in southwest Virginia (Linkins *et al.*, 1990), cellulose disappearance and mass loss were correlated with cellulase activities and that cellulolytic activity declined sharply as the lignocellulose index (LCI) approached a value of 0.7. Ladd and Paul (1973) observed that proteolytic activity increased in soils amended with glucose at rates that coincided with bacterial population growth. Frankenberger and Dick (1983) evaluated relationship between activitie of eleven soil enzymes and microbial respiration, biomass, viable plate counts and soil properties of 10 diverse soils. Alkaline phosphatase, amidase, α-glucosidase and dehygrogenase activities showed significant ($p<0.01$) correlation with microbial respiration in soils that had been amended with glucose. Phosphodiesterase, arylsulphatase, invertase, γ-galactosidase and catalase also revealed significant correlations, whereas acid phosphatase and urease did not show any such relationship. Only phosphodiesterase and α-glucosidase were correlated to microbial numbers ($p<0.05$) when alkaline phosphatase, amidase and catalase were highly correlated ($p<0.01$) with microbial biomass.

Dehydrogenase enzyme systems which are integral part of microorganisms apparently fulfil a significant role in the oxidation of soil organic matter as they transfer hydrogen from substrates to acceptors (Tabatabai, 1994). Because of this attribute the assay of dehydrogenase activity has received considerable attention in past (Skujins, 1967; Klein *et al.*,1971; Ross, 1971). It is also a good index of microbial activity since the average activity of the active population of soil microorganisms is positively and significantly correlated with dehydrogenaseactivity(Skujins,1976).Somestudies(Stevenson,1959) have also shown good correlation between oxygen uptake and dehydrogenase activity but others (Ross, 1973) did not reveal any such relationship. Work by Skuj ins (1973) showed that dehydrogenase activity is highly correlated with carbon dioxide release, proteolytic activit and nitrification potential. Tate and Terry (1980a;b), while studying dehydrogenase activity, reported that when no substrate was supplied exogenously in assay mixture, dehydrogenase provided an estimate of actual microbial respiration under *in situ* conditions.

Activity of different enzymes during the course of detritus decomposition varies and this temporal pattern corresponds to genet patterns of microbial colonization of decaying litter. The species present in early stages of decay lack enzyme systems capable of degrading' cellulose and lignin while species present in latter stages of decomposition have greater capacity to produce celluloytic and lignolytic enzymes (Frankland, 1966; 1969; 1976). A set of studies conducted in north eastern India (Kshattriya *et al.*, 1992; Joshi *et al.*, 1993) revealed such relationship between patterns of forest litter decay and enzyme activities. In both the studies, invertase activity peaked early in litter decay, and declined throughout the subsequent period of decomposition. A similar study by Dilly and Munch (1996) revealed that activities of β-1,4-glucosidase (hydrolyses cellobiose) peaked early and declined steadily during the decomposition of alder litter in northern Germany. As the litter includes compounds like cellobiose and sucrose, the activities of glucosidase and invertase are highest when these compounds are being degraded most rapidly (Moorhead and Sinsabaugh, 2000). Utilization of simple compounds by decomposers is more energetically efficient than using complex polymers, so the degradation of simple compounds precedes that of polymers (Moorhead and Sinsabaugh, 2000). In contrast to invertase activity, Kshattriya *et al.* (1992) and Joshi *et al.* (1993) noted that activity of cellulase and amylase increased more slowly and remained at higher rates for longer periods of time. Likewise, Sinsabaugh *et al.* (1992) studied patterns of enzyme activities associated with the decay of birch wood on a variety of sites in northern New York and found the levels of enzyme activities more closely related to stage of decay than time and generally indifferent to site factors.

Role in aquatic ecosystems

Literature on the activity of enzymes in relation to detritus decomposition is not extensive, particularly, in respect of aquatic ecosystems. However, study of Sinsabaugh *et al.* (1981) reported patterns of β-1,4-exoglucanase, β-1,4-endoglucanase and β-glucosidase during the decomposition of senescent leaves of *Cornus florida, Acer rubrum,* and *Quercus prinus* in a woodland stream. Each enzyme showed a distinct temporal pattern and that the ratio of endoglucanase to exoglucanase activities increased through time and with initial lignin content. Role of pectinolytic enzymes secreted by hyphomycetes, the principal fungal decomposers of plant litter in aquatic ecosystems, in the decomposition of detritus was highlighted by Chamier and Dixon (1982). Significant correlation of cellulolytic and xylanolytic enzymes with mass loss of senescent leaves of *Phragmites communis* in a coastal saline lake was obtained by Tanaka (1991; 1993).

Several other workers have taken a comparative ecosystem approach to litter decomposition in aquatic ecosystems. For example, Kok and Velde (1991) placed litter bags containing fragments of senescent leaves of *Nymphaea alba* in alkaline (pH 8) and acidic (pH 5) freshwater ponds. The contents of each bag were analysed for mass loss, cellulase activity and xylanase activity. In a related study, these workers followed the decomposition of leaf discs of *Nymphaea alba* in six freshwater microcosms with pH values varying from 4.0 to 8.0. Litter from each microcosm was analysed for mass loss and the activities of cellulase, xylanase, polygalacturonase (pectinase), and pectin lyase. These studies pointed out that pectinolytic activity, dependent on pH, is a critical factor underlying differences in mass loss. The activities of phosphatase and five lignocellulose-degrading enzymes were followed by Tank *et al.* (1998) while studying the decomposition of *Liriodendron tulipifera* wood in a small mountain stream from which new inputs of litter were excluded. In comparison to reference stream, fungal biomass, enzyme activities, and breakdown rates were higher in the litter-excluded stream. Variations in the ratios of phosphatase to carbohydrase and phenol oxidase to carbohydrase between the systems indicated that the increased decomposition activity was the result of higher nitrogen and phosphorus availability. Studies of Raviraja *et al.* (1998), Sinsabaugh (1994) and Jackson *et al.* (1995) also confirmed the importance microbial enzyme activities in litter decomposition in aquatic ecosystems.

5.2. Environmental conditions and microbial enzyme activity

Production of extracellular enzymes is affected by all the factors that affect microorganisms in the soil (moisture, temperature, available nutrients) (Kourtev *et al.*, 2002). Once the enzymes are released, their activity is primarily regulated by chemical and physical factors, such as litter chemistry, substrate availability and temperature (Sinsabaugh *et al.*, 1981; Sinsabaugh and Linkins, 1987; Sinsabaugh, 1994). Other factors that influence soil enzyme activities include, organic matter content of soil (Ladd and Buttler, 1972; Dalal, 1975; Speir, 1977), microbial community composition and nature of soil matrix (Chapin et *al.*, 2002). Enzymes involved in the processes that occur only in specific environment (denitrification or methane production and oxidation) are more sensitive to microbial community composition (Gulledge *et al.*, 1997; Schimel, 2001) than enzymes that are widespread and are involved in breakdown of common substrates like proteins and cellulose (Schimel, 2001). Soil enzyme activity is also influenced by the rates at which enzymes are inactivated in soils, either through their degradation by soil proteases or by binding to soil minerals. Binding of an enzyme to the external surface ofroots or microbes frequently prolongs enzyme activity in the soil, whereas binding to mineral particles can alter the enzyme configuration or block the active site of the enzyme, thereby reducing its activity (Chapin *et al.*, 2002). Skujins (1976) noted that seasonal variations in enzymatic activities were generally small; once the enzymes become stabilized in soil they manifest resistance to humidity, temperature and to various environmental changes while studies of some other workers (Khaziyev, 1977; Salfeld and Sochtig, 1977; Dormaar *et al.*, 1984; 1989; Ajwa *et al.*, 1999) showed that there are considerable seasonal fluctuations in enzymatic activities which relate with favourable and unfavourable climatic conditions (Kshattriya *et al.*, 1992; Joshi *et al.*, 1993). Giardina and Ryan (2000) also reported that enzyme activity is limited by temperature only when the supply rate of substrate exceeds the reaction rate for that substrate.

Although decomposition of detritus is largely brought about by extracellular enzymes produced by microbes, yet very limited work is available. In order to fill this lacuna studies need to be planned and executed to workout relationships between activity of various soil enzymes and decomposition under a variety of environmental conditions so as to bring out broad patterns of detritus decomposition.

•••

CHAPTER 6

ABIOTIC FACTORS OF DETRITUS DECOMPOSITION

Decomposition of detritus, in addition to biotic factors, is controlled by different abiotic factors also. Such factors significantly influence decomposition through their effect on the nature of vegetation (substrate quality) and the activity of microorganisms. Specific influence of each of the abiotic factors on microbial activity and litter decomposition is discussed below.

6.1. Soil moisture

Amount and distribution of water within the litter and soil depends on its supply by precipitation, interception by vegetation and loss through evapotranspiration and drainage. Relative importance of these factors is influenced by temperature, physical nature of the soil and characteristics of the vegetation cover. Topography of the area also influences soil moisture by determining lateral runoff. During the periods of heavy rainfall, water may be lost from a soil by drainage under the influence of gravity. However, when precipitation ceases the soil moisture content rapidly stabilizes at the field capacity; value that is fairly consistent for a given soil. Ability to hold water against gravity is largely due to capillary retention in the soil pores. Adsorption of water on the surface of clay and other colloids is also a contributory factor.

Influence on decomposer organisms

Griffin (1963, 1972) has been responsible for focusing attention of the soil microbiologists towards the relationship between matric potential of water in the microenvironment and the activity of microorganisms. Microorganisms absorb moisture directly through their permeable outer covering and must thus directly overcome the increasing tendency of the soil or its other constituents to retain it. Both bacteria and fungi are able to adjust their internal osmotic environment to maintain it hypertonic to the surrounding medium over a very wide range of external potential and it is probably by this means that they are able to maintain their moisture content at about 80%. Growth of fungi by apical extension of hyphae is also dependent on hydrostatic . pressure that is being maintained within the hypha by 'stretching' the freshly synthesized and `soft' apical wall (Swift *et al.*, 1979). Thus, any drop in hydrostatic pressure will affect the growth of microbes. However, ability to adjust to varying water potential of litter and soil during growth is clearly a characteristic of considerable significance to the maintenance of the activity of fungi. Laboratory experiments with microorganisms show that the optimum potential for the apical growth of most fungi lies between -50 and -100 bars; growth of most being severely reduced at the lower end of range. The allowable range for spore germination is usually slightly wider.

In general, an active fungal mycoflora is maintained down to about -150 bars, potential which contrasts strongly with the bacteria which become inactive at potentials below -10 to -15 bars as pointed out by Kouyeas (1964) who followed the growth of soil microbes on microscopic slides buried in soil at different moisture potentials. Bacteria proliferated on the slides only at potentials higher than -1 bar, whilst fungi and actinomycetes persisted beyond -20 bars. Specific bacterial activities, such as nitrification and sulphur oxidation have also been shown to be significantly reduced at potentials of -5 to -10 bars.

Theoretically, Griffin (1972) has shown that mobility of the soil bacteria, either self propelled or passive dispersal by soil water movement, is dependent on the presence of continuous water channels of at least 10 μm thickness. Channels of this diameter will be drained in soils that have potentials below -1 bar. Thus, below this potential bacterial activity will rapidly decline as the substrates providing the food resources at bacterial microsites within the soil are used up without there being an opportunity for relocation on fresh substrate (Swift *et al.*, 1979). Fungi and actinomycetes are not limited in this way as they have the ability to grow across the soil matrix and air filled pores to colonize fresh substrates. Although fungi are active over a wide range of moisture potential, the structure of the fungal community changes markedly in relation to relatively small changes in water potential with zoosporic forms largely evident only above —1 bar (Kouyeas, 1964), while frequency of majority of common moulds declines beyond —50 bars. However, *Aspergillus* and *Penicillium* species may maintain activity down to — 400 bars or beyond. In soils, other than those with very high clay content, fungal activity is only seriously affected below 10% moisture content. Influence of moisture relations on the nature of mycoflora is also seen in standing dead material. Decomposing branches or twigs or more particularly the stems and leaves of herbs and grasses are subjected to extreme desiccation. Webster (1957) and Webster and Dix (1960) have shown that the gradients of moisture content within grass tussocks may be an important determinant of the composition of the fungal flora of decomposing stems and leaves.

Microfauna, particularly protozoa and nematodes, also live within water films and are directly affected by moisture potential. Activity of nematodes is limited at moisture potential of about —1 to —10 bars, probably coinciding with the onset of dehydration. However, many microfauna have adaptations, such as encystment, anabiosis or the production of desiccation resistant eggs that enable them to survive. The resistant spores of fungi and bacteria perform similar function and enable all these microorganisms to respond rapidly to even extreme change in the moisture environment.

In addition to directly affecting microbial activity, several physical processes that can affect microbial activity also vary with soil water content (Paul, 2001). As a consequence, relationship between soil water content and microbial processes in soils is quite complex and depends on the soil moisture retention curve, porosity, concentration of organic matter, pH and soil depth (Goncalves and Carlyle, 1994; Rodrigo *et al.*, 1997; Leiros *et al.*, 1999).

Influence on rate of decomposition

Decomposers, like plants, are most productive under warm moist conditions provided sufficient oxygen is available. This accounts for the high decomposition rate in tropical forests (Gholz *et al.*, 2000) compared to temperate or boreal forests. Decomposition rate of mineral soils declines at soil moisture less than 30-50% of dry mass (Haynes, 1986) due to the reduction in thickness of the moisture films on soil particle surfaces and ultimately due to its influence on the rate of diffusion of substrates to microbes (Stark and Firestone, 1995). Osmotic effects further restrict the activity of soil microbes under conditions of extremely low soil moisture conditions. Bacteria, however, function at lower water availability than do plant roots, hence allow continuance of decomposition in soils that are too dry to support plant activity (Chapin *et al.*, 2002). But under dry or saline conditions high concentration of osmotic metabolites synthesized by microbes creates severe osmotic gradient upon wetting of soil causing many microbial cells to burst. Short term drying-wetting cycles, such as rain storms or the daily dew formation and evaporation can also strongly influence decomposition in litter layer and surface soils. The net effect of drying-wetting cycles, when infrequent, is stimulation of decomposition (as generally occurs in soils), but frequent cycles, as in the litter layer, can reduce microbial population numbers to an extent that decomposition rates may be reduced (Clein and Schimel, 1994). Drying-wetting cycles also tend to stimulate the decomposition of labile substrates (e.g. hemicelluloses), which are broken down largely by rapidly growing bacteria, but retard the decomposition of recalcitrant ones (e.g. lignin) which are broken down by slow growing fungi (Haynes, 1986). Studies (Haynes, 1986) also point towards reduction in decomposition at high soil moisture contents (e.g. >100 to 150% of soil dry mass in mineral soils) because of the impedance of oxygen diffusion into soil. In fact, water acts as an effective barrier to oxygen supply to decomposers in wet soils or in wet microsites within well drained soils because oxygen diffuses 10,000 times more slowly through water than through air and this impeded diffusion of oxygen into wet soils affects rate of decomposition of litter. In a detailed study of the effect of soil moisture potential on decomposition processes, Miller and Johnson (1964) found maximum rate of CO_2 evolution from four soils incubated at 30°C occurred when soil moisture potential ranged from —0.5 to —0.15 bar and the rate at —50 bars was less than 10% of the maximum. The rates of nitrification and total N mineralization showed similar maxima and were negligible below —15 bars.

Wiegert and Evans (1964) also attributed higher decomposition rates of swale vegetation in comparison to upland vegetation to moisture conditions prevalent on swales. Similarly in a Missouri forest, Rochow (1974) reported higher decay rates on the mesic and bottomland plots than on the upland and glade like plots. On the glade like plots the conditions were xeric and the k value for the leaf litter was 0.204.

On the bottomland plots, there was greater availability of moisture and the decomposition rates were higher (k value = 0.568).

In sub-artic conditions with low precipitation, moisture becomes a more important factor than temperature (Karenlampi, 1971). Dry periods are known to significantly affect litter decomposition. Under these climatic conditions, both heavy rainfall and a high percentage of rainy days speed up the weight loss. For example, in a tropical grassland, using mixed grass litter, Gupta and Singh (1977) showed highest disappearance rates of 36.25 to 52.85% from 7th July to 20th October, the period containing maximum number of rainy days.

6.2. Temperature

Temperature varies over the Earth's surface and exhibits a latitudinal graZient modified locally by the aspect of the terrain. Amount of radiations entering the decomposer habitats of litter and soil is also regulated by the extent of vegetation cover (Perfect *et al.*, 1978). Temperature also fluctuates over different time scales-notably seasonally and diurnally within decomposer habitats and these major fluctuations are modified by other features of the litter and soil environment. Whereas the transfer of heat energy in the atmosphere is brought about by radiation and convection, the same transfer within the soil is through the process of conduction. Since soil is a poor conductor of heat, the rate of penetration of heat from the surface into the body of soil is slow. Consequently, diurnal fluctuations of temperature are very marked in the top centimetre of soil and decrease in amplitude with depth. A lag in the response time to surface fluctuations with depth are also noticeable due to the slow rate of conductance of heat by soil. The thermal characteristics of a soil, in fact, result from the features of gaseous, solid and liquid components of the soil. As moisture content is the most variable of these components and as water has a much higher thermal capacity than the components of the solid phase, the heat regime of a soil can vary significantly in relation to its moisture content.

Notwithstanding the diurnal and seasonal variability, temperature plays a key role in regulating decomposition by affecting evaporation rates and microbial activity (Guzman, 1997). As temperature increases, there is an increase in evaporation rate and microbial activity; however, under very high temperatures enzymes responsible for the breakdown of organic matter may start to denature (Becker and Deamer, 1991). Thus, temperature affects decomposition directly by promoting microbial activity and indirectly through its influence on soil moisture and the quantity and quality of organic matter inputs to the soil.

Influence on decomposer organisms

Organisms differ markedly in their response to temperature. Some micro-organisms are clearly adapted to extreme environments and are described variously as thermophiles (adapted to high temperature) or psychrophiles (adapted to low temperature) in comparison with the majority of mesophilic organisms. There have been many suggestions of the temperature limits that should be applied to these groups but it is probable that in nature there is a continuous spectrum of overlapping temperature ranges. However, for a mixed community, high level of activity may be expected over the full range of temperature from about 10°C to 70°C (Doetsch and Cooke, 1973).

Soil animals also show behavioural as well as physiological responses to temperature changes. They seek out favourable thermoclimates by migration, e.g. by vertical movements within the soil and litter. The temperature activity response also differs relative to whether other features of the environment are constant or fluctuating and whether the rate of temperature change is slow or rapid. This latter feature relates to the ability of many animals to acclimatize to temperature change. Thus, whilst all invertebrates are poikilothermic, and theoretically directly responsive to environmental temperature, they do show a range of compensatory mechanisms to maintain a more stable metabolic level. Whilst temperature is a very important regulator of microbial and animal activity, the importance of other environmental and biological factors in modifying the response makes prediction at microhabitat scale a hazardous business. However, responses to temperature at larger scales are readily perceived.

Interesting responses to temperature beyond the range for normal mesophilic activity are observed during composting of plant detritus. The large mass of material in these artificial systems insulates the heap so that the loss of heat to the exterior is much slower than the rate of heat generation by the decomposition process. As a result the internal temperature may reach levels approaching 70-80°C. From this it is clear that high temperatures and period of maximum weight loss and depolymerase activity coincide. This period is also characterized by thermophilic microflora ofbacteria, actinomycetes and fungi with optimal activity typically in the range of 45-55°C. Many of these organisms, particularly fungi, are active decomposers and have the capacity to degrade polysaccharides even at high temperatures. Capacity of the enzymes to remain functionally stable at such temperatures has attracted a lot of attention but the population changes induced by the temperature change may be of greater ecological significance. The shifting population structure of the compost dominated by thermophilic organisms during high temperature period and by mesophiles at lower temperatures may provide a model for the response of decomposer populations to fluctuating temperatures. Data of Perfect *et al.* (1978) indicate that the surface layers of exposed tropical soils may reach temperatures approaching the extreme of the mesophilic range and in fact frequently may go beyond it. Similar high temperatures may occur locally when ambient temperature is high and catabolic activity is generating heat within well insulated microsites. Under such circumstances the thermophilic component of the microflora may be important in maintaining decomposer activity. There is conclusive evidence for population fluctuations between mesophiles and psychrophiles in cold environments as well (Bunnell *et al.*, 1975).

Soil respiration, a measure of microbial activity, increases exponentially over a broad temperature range (Lloyd and Taylor, 1994) due to excessive mineralization of organic carbon to CO_2. While at moderate temperatures most of the respiratory energy supports microbial growth, but at higher temperatures an increasing proportion of energy is used for maintenance and may not lead to a corresponding increase in microbial production. Microbial community composition also changes in response to temperature, usually towards a community dominated by individuals that are adapted and acclimated to higher temperatures. Hence, several physiological and community changes account for the deceptively simple response of microbial respiration to temperature.

Influence on rate of decomposition

Response of all chemical reactions to temperature is very sensitive and enzyme catalysed reactions are no exception. Microbial activity is generally predicted to increase rapidly up to a temperature of about 30°C, with optimal temperature for microbial activity reached between 35°C and 45°C (Paul and Clark, 1996). Although all biological reactions have temperature optima, this is neglected in most models because even average daily temperatures within the surface 5 cm of soil are unlikely to reach this optimum. Furthermore, in the field such extreme temperatures generally coincide with periods of moisture limitation because of which decomposition is largely unaffected by temperature (Paul, 2001). Consequently, the combined effect of temperature and moisture is more prominent than the effect of temperature alone. Jenny *et al.* (1949) reported that the residues of alfalfa leaves were smaller at high temperatures and moist conditions under the tropical climate in Colombia and Costa Rica and were larger at low temperatures in the Sierra Nevada mountains in California (Witkamp and van der Drift, 1961). Such differences, as stated earlier, influence the rate of decomposition.

Temperature may also affect decomposition through freeze-thaw events. Freezing kills many of the microbes present in decomposing litter and soil organic matter thereby releasing soluble organic material into the soil. This pulse of available substrate can support rapid decomposition and nitrogen mineralization with the onset of favourable conditions (Lipson *et al.*, 1999). Freezing and thawing also stimulates decomposition by physically disrupting soil aggregates and the cellular structure of litter, thereby exposing fresh surfaces to decomposition. It is because of it that in some arctic ecosystems the decomposition that occurs during autumn, winter and spring accounts for most of the annual litter mass loss (Hobbie and Chapin, 1996).

Temperature has many indirect effects on decomposition as well. For example, high temperature reduces soil moisture by increasing evaporation and transpiration (Guzman, 1997) and the consequent soil drying reduces decomposition in dry climates. Stimulation of microbial activity by warm temperatures also initiates a series of feedback loops, such as the consumption of oxygen in microbial and root respiration that influence decomposition. On the other hand, nutrients released by decomposition at high temperature increase the quantity and quality of litter produced by plants, thus altering the substrates available for decomposition. High temperatures also increase the rate of chemical weathering, which in the short term enhances nutrient supply. In cold climates, low temperature leads to a layer of permanently frozen soils (permafrost) that restricts drainage and, consequently influences decomposition. It has been hypothesized that the positive response to increases in temperature is higher for decomposition than for primary productivity (Woodwell, 1978; Jenkinson *et al*., 1991; Schimel *et al*., 1994; Kirschbaum, 1995). Analysis by Post *et al*. (1982) shows that increasing temperatures increase the rate of soil carbon output more than the input, which implies that the temperature response function for decomposition is steeper than for production. Hence response of soil organic matter to increasing temperature is one of the most important likely changes with climate change (Agren *et al*., 1996).

As stated earlier, temperature is one of the basic tenets of the decomposition kinetics theory which has been well described by several authors (Bunnell and Tait, 1974; Swift *et al*., 1979). Strong correlation between soil respiration and temperature, noted first by Lundegardh (1927), has been quantified for many soils under different conditions (Singh and Gupta, 1977; Raich and Schlesinger, 1992; Lloyd and Taylor, 1994; Kirschbaum, 1995). However, there is no unanimity among researchers on the form of the relationship between decomposition and temperature (Katterer *et al*., 1998). In many decomposition studies, the temperature coefficient, Q_{10} relationship (van't Hoff, 1898) is used to describe the difference in decomposition vis-à-vis temperature (Swift *et al*., 1979; Katterer *et al*., 1998):

$$k_2 = k_1 Q_{10}^{(T_2-T_1)/10}$$

where k_2 and k_1 are the rate constants at two observed temperatures of T_2 and T_1. Kirschbaum (1995) reviewed 20 different data sets and compiled reported Q_{10} values. Fitting the Q 10 function to this data set resulted in high Q_{10} values at low temperatures decreasing from about 8 at 0°C to 2.5 at 20°C.

For biological systems, Q_{10} is assumed to be approximately 2.0 but deviations from this are fairly common (Swift *et al*., 1979). This is due to both the changing nature of the response as the limits of activity are approached and other environmental and biological features that vary with temperature and may limit the biological response. This Q_{10} relationship is easy to use and provides a ready indication of the temperature sensitivity of any system but it lacks a theoretical justification (Agren *et al.*, 1996) and is not constant for any system (Kirschbaum, 1995). Thus, apart from Q_{10} relationship, several other functions have been used to describe temperature responses, like linear functions (Witkamp, 1966; Froment, 1972; Gupta and Singh, 1981b), powerfunctions(KuceraandKirkham,1971),Arrhenius—typefunction (HowardandHoward,1979;EllertandBettany,1992;LloydandTaylor, 1994) and the heat sum concept (Andren and Paustian, 1987; Honeycutt *et al.*, 1988). These functions can easily be converted into each other by changing parameters into temperature dependent variables. Thus, a comparison between studies where different approaches have been used is possible.

6.3. pH

Negative logarithm of the activity of H^+ in solution is denoted by pH and it indicates the acidity /alkalinity of soil when measured in an aqueous or KC1 extract (Subba Rao, 2000). In the latter case, some of the H^+ ions adsorbed to soil colloids are also replaced by k^+ and hence the pH indicated is lower than water extract. The buffering capacity of soil, depending on the type of soil, also determines soil pH. Soils with high organic or inorganic colloids exert greater buffering capacity. pH value of soils varies from 3-10 (Subba Rao, 2000); the acid sulphate and podzolic soils are low in pH whereas calcareous and alkali soils are very high in pH.

Role of pH is among the most difficult of environmental factors to understand but is of central importance (Brady and Weil, 2001). The difficulty lies partly in the complex interaction of other factors (such as the nature and size distribution of particulate phase, the concentration of cations and their precipitation) which determine soil pH and partly in the complexity of biological and chemical effects. Swift *et al.* (1979) pointed out that information available for mineral layer of soil regarding pH is far more than for litter but activity of decomposers in the organic layer is an important determinant of the pH regime in the underlying soil.

pH of a soil, or soil horizon, is generally represented by a single value, or a range of values, based on standard measurements. However, there are marked variations in pH between microsites within a soil sample and these variations are ecologically very important. So, bulk pH (pH_B) measurement is made which is an average of all the local pH values (pH_1) and is useful for comparing different soils or horizons. Bulk pH is determined primarily by the balance between hydrogen ions (and to some extent Al^{3+}) and basic cations, such as Ca^2 +, K^+ and Me^{2+} This balance is modified by other environmental factors like the nature of the solid matrix of soil and precipitation. Significance of solid matrix lies in the proportion of negatively charged sites which it contains. Among the mineral fractions, clays have charged surfaces because of unsatisfied valences. Many organic colloids of the humus fractions are also negatively charged. Presence of the negative charge at the surface of these materials leads them to attract a 'cloud' of cations to the surface region.

In acid soils, H^+ ions tend to predominate at the colloid surfaces but in base rich alkaline or neutral soils the main components may be basic cations like Ca^{2+}. These states are by no means permanent. The hydrogen ions are adsorbed more strongly than other cations so that when an excess of H^+ are produced (as during the formation of organic or mineral acids, such as H_2CO_3 by the decomposition of organic matter) they will exchange with the Ca^{2+} or other basic cations on the surface.

$$Ca\ (adsorbed) + 2H_2CO_3 \longleftrightarrow 2H^+\ (adsorbed) + Ca\ (HCO_3)_2$$

This equilibrium may shift to left if excess of Ca 2^+ occurs in soil solution due to its release from $CaCO_3$. However, if leaching occurs then the removal of soluble calcium bicarbonate throws the equilibrium strongly to the right and leads to build up of hydrogen ions on the clay surfaces.

Bulk pH of litter or other detritus is usually determined by measuring the pH of a liquid in equilibrium with either intact or macerated resource material. The pH of plant material is usually acidic; leaves of temperate deciduous trees generally show a pH range of 5.0-6.5 while conifer needles have pH ranging from 3.5-4.2 (Broadfoot and Pierre, 1939). The pH of wood is 0.5-1.0 lower than that of leaves with conifers again having lower pH than hardwoods (Campbell and Bryant, 1941).

The cell wall components of plant materials may exhibit a considerable degree of cation exchange behaviour in a manner analogous to that of soil colloids. Impure cellulosic products possess a net negative charge which is attributed mainly to the pectic acids and lignin components but partially modified celluloses may also have this property due to the presence of carboxyl groups. The negative charges on these polymers attract cations, which in acid conditions exchange with hydrogen ions in obedience to the mass action law (Sookne and Harris, 1954). Thus, local differences in pH occur within decomposing litter.

The products of decomposition are predominantly acidic but unless the local pH change is extreme the plant materials will have a tendency to retain a high proportion of basic cations. The rate of loss of cations from decomposing litter shows an inverse relationship with their affinities for charged fibres (i.e., Ca^{2+} binds more strongly than Mg^{2+}, K^+ or Na^+). Plant litters thus remain alkaline in comparison to the highly fragmented materials of the fermentation and humus layer. In fact the generalization has been made that the initial impact of decomposition is to cause an increase in the bulk pH of plant materials. Sjors (1959) found an average rise of 0.7 pH units from a starting mean of pH 5.7 for twelve hardwood leaf litters after one month's decomposition. Decrease in H^+ concentration implied by this is usually attributed to the leaching out of acidic material, particularly the component of the vacuolar sap. Nykvist (1963) found the highest acidity and organic acid content in leachates during the earliest stages of decay and Frankland *et al.* (1963) have shown a strong correlation between rainfall and pH change in decomposing litter.

Most soils show greatest acidity in the O or A horizons which contain the products of primary decomposition and at this stage of decay the formation of acidic products is high and mineralization proceeds at such a rate that the buffering effect of bound cations is destroyed. Below the immediate 'decomposer horizon', however, the buffering capacity of most soils is such that the leaching of acids from litter has no noticeable effect on bulk pH, as in mineral soils (Frankland *et al.*, 1963). Thus, decomposition processes help to determine the pH of environment that in turn regulates the activity of decomposition processes.

Micro-flora and micro-fauna, living within the water films at the surface of charged particles in soil, are most susdeptible of decomposer organisms to the effects of variation in local pH though some of them have the capacity to modify the pH in their immediate environment.

The microorganisms grown in pure culture bring about a pH shift in the medium, usually resulting in increased acidity. This shift has been attributed to selective uptake of cations (e.g. of NH_4^+ in preference to NO_3-) or production of organic acids and CO_2. An extreme example is the action of the sulphur oxidizing bacterium *Thiobacillus thiooxidans* that generates sulphuric acid in its immediate environment and can readily flourish even at pH of 1.0. Conversely, the production of ammonia during proteolytic activity can create an environment with decreased acidity. The local pH around microbial cells can also be influenced by the negatively charged cell surface (Marshall, 1971), which may attract a local concentration of H^+ in a manner analogous to that of clay minerals.

There is a large body of information detailing the effect of pH on various aspects of the physiology of bacteria, fungi and invertebrate animals (Dickinson and Pugh, 1974; Agren *et al.*, 1996; Dalal, 2001; Chapin *et al.*, 2002). Some broad generalizations that can be drawn are: bacteria tolerate relatively narrow ranges of pH at the alkaline end of the spectrum.; fungi have generally broad optima but are most active at acidic pH. Physiological basis of such effects are probably complex and a variety of different mechanisms operate. In bacteria, pH of the cytoplasm closely mirrors the external medium but fungi seem to be able to regulate their internal pH and maintain it between 5 and 6. Same is true of most invertebrates where the digestive system is in most cases maintained at a weakly alkaline pH.

Effects of pH on extracellular enzymes may be indirect. As proteins are amphoteric in nature, they have very specific pH optima, although some can operate over a broad range. But there is little information on the pH spectra of the extracellular depolymerising enzymes of decomposers. McLaren (1960; 1962) has pointed out that considerable differences occur in behaviour of enzymes when particles or other solid surfaces are included in the systems because enzyme molecules have a high probability of adsorption onto charged surfaces which means that the pH and ionic environment in which they act is different from that of the soil solution. Other workers also have demonstrated, for a number of different enzymes (proteases, phosphatases, urease), that the optimum pH for these enzymes is higher when adsorbed on clay minerals than in solution, presumably because of a decreased pH at the adsorbent surface (e.g. Skujins, 1967). Adsorbance has also been shown to affect the persistence of enzymes in the soil (e.g. their resistance to proteolysis). Substrate molecules may also adsorb to charged particles and although local pH is only one of the number of factors involved, this can result in markedly different rates of catabolism when compared with controls (lacking particles). Hence the responses are complex and both stimulation and inhibition of the rate of mineralization have been observed (Stotzky, 1974).

There are also many reports of differences in the activity of microorganisms between sorbed and non-sorbed sites and some of these have been attributed to local pH effects. A much-quoted example is that of nitrification. McLaren and Skujins (1963) showed that Nitrobacter agilis had a pH optimum of about 6.0 in liquid suspension for production of NO_3^- from NO_2^- but this increased to 6.5 or 7.0 in presence of soil colloids or ion exchange resins. They interpreted this in terms of the repulsion of NO_2^- ions by the negatively charged surfaces which at the same time accumulated H^+ thus creating a decreased pH_L . Decreasing the H^+ concentration in the solution raised the pH_L at the surface region and also permited ready aces of the NO_2^-. Thus, in this particular case pH change is not so important but the key factor in determining the rate of reaction is pH equilibrium.

Furthermore, pH of the soil solution may affect the availability of essential elements to decomposers because of its influence on their solubility and ionic form. The most dramatic effect is on iron, which precipitates out as insoluble hydroxide from ferric salts at pH 3 and from ferrous salts at pH 5. In general, more acidic conditions increasethe solubility of a wick range of necessary elements including P, Ca and K. In soils where these elements exist largely in insoluble form, the influence of pH in making them accessible may be of great significance.

From the foregoing discussion it becomes evident that the effect of pH on the microflora and fauna is often subtle and expressed at a local level. The meso- and macro-fauna are largely able to avoid such effects both because of the scale of their habitats and by possession of an external integument. Nonetheless pH_B can often be distinguished as a factor determining the distribution and activity of these organisms. In this respect, earthworms have been studied more intensively than any other group. The soft bodied nature of these organisms makes them relatively sensitive to contact with solutions of varying acidity and as a result they show marked distributional discontinuity in relation to pH. Bornebusch (1930) found that forest soils with pH 4.3 or less in Denmark contained few earthworms while neutral or basic soils supported the largest worm populations. Bornebusch's results were supported by Satchell (1955, 1967) who classified the British Lumbricidae according to their pH tolerance ranges (Satchell, 1955; Edwards and Lofty, 1977).

All else being equal, decomposition occurs more rapidly in neutral than in acidic soils due to a variety of interacting factors including changes in plant species composition and associated changes in quantity and quality of litter.

6.4. Fire

Fire plays a major role in modifying the physical, chemical and biological characteristics of soil, particularly grasslands that occupy one third of the Earth's surface (Viro, 1974; Senthilkumar *et al.*, 1997). Fire, a natural disturbance component of certain ecosystems, results in the changes in soil temperature (microclimate), water potential, plant species composition, nutrient and microbial status (Knapp and Seastedt, 1986; Gibson and Hulbert, 1987; Gibson and Hetrick, 1988; Collins and Gibson, 1990). Effect of burning and its influence on nutrient cycling has been examined in several ecosystems and such studies have generally pointed towards shrub or tree cover developing during the fallow period (Jorgensen and Hodges, 1970; Amaranthus and Trappe, 1993). In general, the burning process releases nutrients to soil and large amounts of C and N into the atmosphere which are otherwise locked. Even though a large number of studies on the influence of fire on soil physical and chemical properties are available, little information is available on its impact on soil biological properties (Perry *et al.*, 1987; Senthilkumar *et al.*, 1995). The build up of plant secondary compounds, such as monoterpenoids may increase the severity of a fire and volatilise nutrients in the soil (Overby and Perry, 1996). On the whole biotic activity increases following a fire (Anderson and Menges, 1997), which is dueto rapid decomposition of organic matter (Guzman, 1997).

Effects of fire on soil organic matter (SOM) are highly complex and differ greatly among different sites with different soils, climates, species composition, and time scales (Viro, 1974; Walker *et al.*, 1986; Raison *et al.*, 1989). Fire potentially has three effects on SOM (Agren *et al.*, 1996):

a) it causes a short term loss of C directly due to combustion and indirectly due to changes in micro-climate that enhance decomposition rate,
b) it causes a short to medium term loss of N and a long term loss of P and other nutrients, and,
c) it may change species composition of a site that may restore lost N.

Fire leads to direct combustion of large amounts of C above the soil surface and the combustion may be almost complete for dry grasslands or may consume only a small fraction of potentially available C, if the intensity of fire is low particularly in an established and fire resistant forest (Walker *et al.*, 1986). Intense fires may also burn part of the organic matter that is already in the soil (Walker *et al.*, 1986), but soils tend to heat up only very slowly and even very intense fires are unlikely to directly affect the SOM to a depth greater than a few centimetres (Aston and Gill, 1976; Humphreys and Craig, 1981).

Post-fire changes in the soil micro-climate are also important. After the canopy and understory have been burned off, the soil is no longer protected by shade and much more sunlight may reach the soil surface. This can raise both maximum and minimum soil temperature by several degrees for one or more years (Armson, 1979). Viro (1974) reported an increase in mean summer temperature in the year after the fire in the humus layer from 18°C under the unburned tree canopy to 31°C in the open. With the loss of leaves from vegetation, the water transpiring capacity is also reduced and more water is retained in the soil profile. Hence, for one to several years after a fire soils may be warmer and wetter than in unburned state, which in turn may enhance the rate of organic matter decomposition. It would be pertinent to mention here that decrease in the water holding capacity due to loss of organic matter together with increased soil temperature may lead to increase in evaporation from the soil (Pritchett, 1979) with implications for litter decomposition.

While fire leads to loss of soil C in the short term, the longer term C dynamics is principally controlled by the soil's nutrient economy. During fires, N compounds readily volatilise and consequently large quantities may be lost (Raison *et al.*, 1985). Viro (1974) estimated the average loss of N due to burning after clear felling to be 320 kg ha^{-1} with about 180 kg coming from slash and 140 kg from humus layer. Phosphorus compounds, on the other hand, are generally less volatile and a greater proportion of P is likely to be retained within the system (Raison *et al.*, 1985), although ash deposited on soil surface may readily be lost due to erosion (Raison *et al.*, 1989). Another consequence of fires can be a substantial increase in soil pH by 2-3 pH units in acid soils which, in turn, can enhance N fixation and mineralization (Johansson, 1984) and the amount of available nutrients (Nykvist, 1977). If the nutrients associated with the lost C become available to plants and promote greater C gain, then after a number of years the soil C to nutrient ratios would return to the values characteristic for the site and the same soil C content as before the occurrence of fire would be re-established. The direct loss of C due to fire, hence would have few long term consequences. In regions where fire frequency and intensity may increase more nutrients would be lost and overall productivity and soil C storage may eventually decrease. The converse may be true for regions with fewer fires (Agren *et al.*,1996). Notwithstanding the deleterious influences of fires, prescribed burning has been suggested as a useful tool for opening up shrub thickets or triggering sprout reproduction (Severson and Boldt, 1978; Ram and Ramakrishnan, 1988).

Effect of fire on soil biochemical properties is not clearly understood. The action of both heat and ash modify the soil chemical and physical properties (Viro, 1974), and alteration in the microbiological activities may also be expected. Assessment of soil enzyme activities has been suggested to provide a measure of microbial activity in fire disturbed soils (Ladd, 1972) and since soil enzymes are involved in various decomposition and chemical transformations, the measurement of enzyme activities can give indications about the extent of specific processes involved in governing soil fertility (Mishra and Pradhan, 1987). Prescribed fire is considered as a physical treatment which may change soil enzyme activity when soil is partially warmed by heat. It has been shown to be conducive for the colonization by soil microbes as nutrients are liberated from the biomass and competition is temporarily arrested (Tu, 1982). Senthilkumar *et al.* (1997) estimated amylase, cellulase, invertase and phosphatase activities in grasslands and reported that burning stimulates soil enzyme activities in the surface layer. A two-fold increase was recorded in amylase activity while other enzymes showed relatively small increase. This increase was attributed to the increase in the population of soil microorganisms brought about by the relative increase in nutrients and organic matter recorded in the burned area together with favourable moisture and temperature following grass fire (Senthilkumar *et al.*, 1995). Similar factors have been indicated by other workers (Meiklejohn, 1955; Ahlgren, 1974) as responsible for increase in microbial activity after burning. Although partial sterilization occurs in soil after fire (Ahlgren, 1974), rapid colonization by microorganisms present in the air, water or unburnt soil takes place. Apart from stimulatory effect of burning on soil enzyme activity, 10-25 fold increase in microbial population has been reported (Ahlgren and Ahlgren, 1965; Vazquez *et al.*, 1993). Studies carried out by Dhillion and Anderson (1993) revealed that roots of plants on burnt sites supported more fungal biomass than roots on the unburnt sites, suggesting the possibility of increased secretion of root exudates after fire. It has also been shown that the burned soils and plants contained more nutrients than plants in unburned plots (Dhillion and Anderson, 1993). The elevated microbial biomass and increased nutrient availability from ash and other carbonaceous material derived from plants thus result in elevated soil enzyme activities in grasslands. Ross (1973) has shown that soil respiratory activities decrease with depth and invertase activities are generally high in grasslands following fire. Senthilkumar (1995) reported higher biomass turnover in grasslands than forest or agricultural soils and biomass production was estimated to be in the range of 5.3-7.0 k gm^{-2} $year^{-1}$ which is on the higher range of values reported for tropical and temperate grasslands. Therefore, a high rate of enzyme activity is expected in grasslands when the conditions are favourable for the growth of microorganisms.

Wildfire increases air temperature up to 85-840°C while increases in soil temperatures are usually less than 50-80°C and are restricted to the top 3-4 cm of soil and persist for only few minutes (Benthley and Fenner, 1958). Activities of enzymes involved in carbon, nitrogen and phosphorus cycles have been shown to increase with an increase in temperature from 10 to 20°C (Frank and Malkomes, 1993). Nannipieri *et al.* (1982) showed that a slight increase in soil temperature raised the activities of phosphatase, urease and protease. Besides, combustion products (fly ash) and burnt materials (bed ash) influence soil enzyme activities favourably (McCarty *et al.*, 1994), similar to that produced by soil amendments, such as $CaCO_3$. Thus, heat and combustion products of a grass fire may have little direct effect on soil microbial activities.

6.5. Soil atmosphere

Composition of the atmosphere in decomposer environment is a product of both biological and physical phenomena. Under aerobic conditions oxygen is used up and carbon dioxide released, resulting in a progressive reversal of the normal atmospheric ratio of the two gases. Under anaerobic conditions the gases other than oxygen and carbon dioxide, e.g., methane and hydrogen sulphide, begin to predominate. Besides, the composition of the gaseous environment within litter and soil is largely determined by factors that control the diffusion rate between pores and external atmosphere and in this respect the intervention of the liquid phase is of particular importance. Major gases in soil are nitrogen, oxygen and carbon dioxide. Carbon dioxide content of soil air (0.3-1.0% by volume) is higher than that of the atmospheric air (0.03% by volume) (Subba Rao, 2000). This is due to consumption of oxygen during respiratory activity of the soil.

Oxygen and carbon dioxide differ markedly in their properties and in their effects on decomposer organisms. However, the interactive effects between the two may be more significant than those of either molecule alone. The highest levels of decomposer activity are found in aerobic environment; anaerobiosis usually results in incomplete degradation of substrates and accumulation of the organic matter. However, this distinction into aerobes and anaerobes is misleading as there exists a continuous range of relationships from obligate aerobes through facultative forms to obligate anaerobes. However, fungi and most meso- and macro-fauna though strict aerobes also have the capacity to survive under anaerobic conditions as well, while in contrast many bacteria and some nematodes and protozoa are active in fully anaerobic conditions. Many facultative anaerobes utilize fermentative pathways even in presence of relatively high levels of oxygen. Greenwood (1968) showed that the switch from aerobic to anaerobic respiration in the soil occurs at an O_2 concentration of about 3 μM. This is about one hundred times higher than the Michaelis constant for saturation of terminal oxidases of aerobic respiration and it is indicative of the operation of limiting factors other than just the concentration of oxygen in the soil atmosphere. Thus, changing oxidative conditions produce a varying spectrum of response in the decomposer community. Nevertheless, decomposition in anaerobic environments occurs slowly and produces energy inefficiently as oxygen is the preferred electron acceptor. When available, it provides the highest energy return per unit of organic matter oxidized. Wetlands, estuaries and sediments beneath lakes and oceans which occupy vast areas of the Earth's surface are such environments where oxygen supply limits decomposition rate. In these habitats organisms must use other electron acceptors to derive energy from organic matter. Anaerobic organisms are known to exist even in the upper horizons of apparently well aerated soils and it has been postulated that local pockets of anerobiosis occur in otherwise aerobic habitats. Basis for this is the poor solubility and slow rate of diffusion of O_2 in water. As most microorganisms particularly bacteria exist within the water phase in soil, the continued oxygenation of their sites depends on the diffusion of O_2 through the water film. If the pore space of a soil is substantially filled by water, the extent of oxygenation at the surface of soil crumbs may be negligible. Even in much drier soils the oxygenation may be limited by moisture content. Greenwood (1968) calculated, on the basis of normal microbial respiration rates, that the rate of diffusion of oxygen in water would be insufficient to replace oxygen lost over distances greater than 3 mm. Thus, the centre of crumbs of greater radius than this would be expected to be anaerobic even in well aerated soils. Substantial suppression of aerobic decomposition in soil may occur during times of water logging. In many soils this may be only a temporary state but the extensive accumulation of organic matter in areas of high rainfall and poor drainage is largely attributed to this mechanism.

Within soils, oxygen content varies with depth, season and type of soil. Very little information is available for litter layers but Brierley (1955) showed that the oxygen content in beech litter as deep as 15 cm did not fall below 19.5%. Oxygen depletion is probably more common in relatively massive resources, such as branches, stems or roots of trees (Paim and Beckel, 1963). Paim and Beckel (1963) reported that it is likely that inhibitory or even anaerobic oxygen levels many occasionally prevail in decaying wood, particularly, if the water content is high. Most available evidence suggests little decrease in the activity of wood decay organisms above oxygen levels of about 2%. Below this level, Paim and Beckel (1963) could find no evidence of invasion by cerambycid beetle larvae, suggesting that wood decaying fungi may be more tolerant of oxygen depletion. Lopez-Real and Swift (1977) showed that hyphal growth of fungi within wood blocks could occur when they are in equilibrium with an external atmosphere containing only 0.5% oxygen. When the blocks were placed in an atmosphere of N_2, fungal activity ceased after a short interval of time indicating that inhibition occurs at the time of total consumption of oxygen i.e., with the onset of anaerobic conditions. Griffin (1972) reviewed the response of soil fungi, measured in terms of their germination or linear extension growth rates, to varying concentrations of oxygen and concluded that most species are unaffected by reduction of the oxygen partial pressure in the surrounding gaseous atmosphere to below 4%.

In contrast to oxygen, dynamics of carbon dioxide within decomposer habitats is markedly different as CO_2 is about 35 to 55 times more soluble in water than oxygen over the range of temperature common in soils. This means that higher moisture contents do not affect the rate of diffusion of CO_2 in litter and soil. Another important characteristic of CO_2 is its tendency to dissociate in water into a number of different ionic forms as shown below:

$$CO_2\text{ (atm)} \leftrightarrow CO_2\text{ (dissolved)} + H_2O \leftrightarrow H_2CO_3$$
$$H_2CO_3 \leftrightarrow H^+ + HCO_3^- \leftrightarrow H^+ + CO_3^{2}$$

Relative frequency of these forms is also affected by pH. Solubility and the ionic equilibrium are affected by the presence of other ions and by the temperature. It thus becomes clear that for organisms, such as fungi and bacteria, measurement of pH along with gaseous CO_2 is also important for interpretation of the prevailing environment. Griffm (1972) concluded that bicarbonate is the most significant form in respect of fungi; consequently the response of microorganisms to changes in the partial pressure of CO_2 would be greater in alkaline environments (where bicarbonate ion would be more likely to be present at biologically significant concentrations) than in acidic environments. For soil animals inhabiting air filled pores only, the partial pressure of gaseous CO_2 is of environmental significance because high levels of CO_2 are known to have a narcotic effect on most insects but there is evidence that many soil invertebrates are adapted to withstand higher than average levels.

Whether or not the global increases in atmospheric CO_2 would result in net terrestrial ecosystem carbon storage is still being actively debated (Melillo *et al.*, 1996). The average annual rate of increase of CO_2 is 1.8 ppm with a predicted doubling of pre-industrial concentrations by the end of the 21St century (Houghton *et al.*, 1996). Despite an increasing number of studies on soil processes (Canadell *et al.*, 1996) little is known about the overall impact of elevated atmospheric CO_2 onet ecosystem C storage (Amthor and Koch, 1996). Eforts are being directed towards studies which examine vegetational response to increased CO_2 levels (Roberts, 1989; Henderson *et al.*, 1993; Coffin and Lauenroth, 1996; Somartne and Dhanapalal, 1996). These studies suggest that vegetation distributions will exhibit a response of shifting their distribution patterns (Guzman, 1997). Couteaux *et al.* (1995) have hypothesized that there will be enhanced decomposition rates and a change in chemical quality of litter, especially in nitrogen concentration. Findings of Hungate *et al.* (1997), who studied effect of increased CO_2 in a grassland, reported an increase in carbon cycling and a small accumulation of litter in the grassland. On a long term basis, litter will slowly increase up to a point where there would be a negative feedback in the system in the form of increased surface litter which contains nutrients like nitrogen. If this process of litter accumulation continues, fire would be the main agent of litter decomposition in grasslands (Guzman, 1997). Fire volatilises a large portion of organic matter and also adds some of this material back into the soil (Kutiel and Inbar, 1993). Furthermore, it has been demonstrated that elevated CO_2 favours investment of biomass in roots (and their exudates) relative to that in leaves (Stulen and Den Hertog, 1993) especially when plant growth is nutrient limited. Studies of van Ginkel and Gorissen (1998) revealed that elevated CO_2 significantly increased shoot biomass by an average of 28%, while root biomass increased by an average of 42 % in rye grass at both low and high soil nitrogen concentration. Hence, the suggestion that rise in CO_2 will lead to increase in plant biomass, and hence terrestrial C storage is an oversimplification, since belowground C storage would dominate in some terrestrial ecosystems (Anderson, 1992). C storage in ecosystems (i.e., net ecosystem productivity, NEP) is the difference between net primary production (NPP) and ecosystem heterotrophic respiration both of which could change as atmospheric CO_2 concentrations increase (Norby *et al.*, 2001). Study of van Ginkel *et al.* (1996) on decomposition of roots of Lolium perenne under elevated CO_2, revealed a higher root yield but lower decomposition at elevated CO_2 which in turn led to longer residence time of carbon in soil and also to higher carbon storage. Therefore, the most pronounced plant response to elevated atmospheric CO_2 would be accumulation of below-ground biomass (Rogers *et al.*, 1994). A related study was conducted by Nitschelm *et al.* (1997), who studied effects of atmospheric CO_2 enrichment on soil organic carbon content and noted that the occurrence of both greater plant material input through higher yields and reduced residue decomposition rates at elevated CO_2 would be expected to impact soil C storage significantly. They stated that soil carbon input and storage under elevated CO_2 tend to increase and hence soils may be important sinks for increasing atmospheric CO_2 (Ginkel and Gorissen, 1998). Thus, with increase in the atmospheric CO_2 concentration it is likely that temperate grasslands will sequester greater amount of C for longer periods of time. This phenomenon would significantly increase the carbon sink strength of these ecosystems.

Two general issues of concern are whether increases in NPP in response to rising CO_2 will be sustained if negative feedbacks through N-cycle occur, and whether decomposition rates will change if atmospheric CO_2 concentration affects the chemistry of the substrates for decomposition. These issues in turn depend on chemistry of litter under elevated CO_2 conditions (Norby *et al.*, 2001). The central role of both N and lignin in controlling litter decomposition rates (Melillo *et al.*, 1982; Fog, 1988; Taylor *et al.*, 1989) suggest that any change in these determinants of litter quality could lead to subsequent alterations in litter decomposition. For example, reduction in N content (with associated increase in C:N ratio) of litter within a species would be expected to result in decreased rate of litter decomposition regardless of the atmospheric CO_2 (Cotrufo *et al.*, 1995) and lignin would assume importance in later stages of decomposition (Berg and Staaf, 1980; Berg *et al.*, 1987). Though degradation of lignin is accomplished by a limited group of soil microorganisms, it is known that high N levels may suppress lignin degradation rates (Berg and Ekbohm, 1991; Fog, 1988; Cotrufo *et al.*, 1998a and b). Thus, any decrease in litter N occurring under elevated CO_2 will lead to an increase in lignin degradation in the final stages of decomposition (Norby *et al.*, 2001). Decomposition is also a controller of N availability (Marrs *et al.*, 1983). Torbert *et al.* (1998) carried out studies on crop residue decomposition under elevated atmospheric CO_2 and pointed out that as the CO_2 content of air rises, even higher plants will increase their photosynthetic rates, thereby removing more carbon from the atmosphere and integrating it into their tissues eventually resulting in higher C:N ratios. Following senescence, leaves and stems typically fall to the ground and become incorporated into the soil, where the mineralization of their various elements occurs. However, the amount of carbon mineralization may decrease as the CO_2 content in air increases, thereby locking up greater stores of carbon in the soil and helping to slow the rate of rise of the CO_2 content in the atmosphere.

Long term ecosystem responses to elevated CO_2 may ultimately depend on N availability to plants and its efficient utilization under conditions of increased atmospheric CO_2 (Norby *et al.*, 2001). The extent to which the C:N ratio of plant tissues can change is one of the key biogeochemical determinants regulating the amount of C that can be sequestered from atmosphere into vegetation (Rastetter *et al.*, 1992). The possibility that the carbon dioxide enrichment will alter the chemistry of litter in a way that would lead to reduced decomposition and N availability was first proposed by Strain and Bazzaz (1983). At that time there was evidence indicating lower foliar N concentration in CO_2 enriched plants and it was reasonable to asume thathe N concentration in leaf litter would also be lower. This, in turn, would result in lower N availability in the soil, setting in motion a negative feedback to photosynthesis and plant growth. This scenario, called the "litter quality" hypothesis has been frequently invoked to explain increases in growth of plants observed in short-term experiments under elevated CO2 *et al.*, 2001). In fact, several compilations of the literature have (Norby reported declines in foliar N concentration (O'Neill and Norby, 1996) with tree species averaging 21% (Mc Guire *et al.*, 1995), 16% (Curtis and Wang, 1998), 14% (Cotrufo *et al.*, 1998c), or 13% (Norby *et al.*, 1999). Much of this decline is related to an increased dry matter content of leaves rather than reduced total N content (Curtis, 1996; Norby *et al.*, 1999), but it is the C:N ratio rather than total N content that is more important expression of litter chemistry with regard to decomposition (Norby *et al.*, 2001). Results of CO_2 enrichment experiments have been mixed with some studies indicating lower N (reduced litter quality) (Cotrufo et al., 1994; Jongen *et al.*, 1995; van Ginkel *et al.*, 1996) and others indicating no effect of CO_2 on litter quality (Gahrooee, 1998; Hirschel *et al.*, 1997; O'Neill and Norby, 1996; Randlett *et al.*, 1996).

The relationship between N in green leaves and N in litter is controlled by the process of resorption, which involves the breakdown of proteins during leaf senescence and amino acid translocation in the phloem to perennial tissues. Nitrogen resorption is a proximate controller of leaf litter N, so any consideration of CO_2 effects on litter chemistry should explicitly consider CO_2 effects on resorption (Norby *et al.*, 2000). Although there is wide variation among species in resorption efficiency, or the fraction of green leaf N that is resorbed during senescence (Chapin and Kedrowski, 1983), most ecosystem models set resorption efficiency as a constant, typically 50%, which is the average value across many species (Aerts, 1996, Chapin and Kedrowski, 1983).

It is now apparent that rise in CO_2 will not consistently decrease litter quality across species or ecosystems (Norby *et al.*, 2001) although changes in litter chemistry of individual species are likely to affect litter decomposition rates of some ecosystems (Cotrufo *et al.*, 1994; Cotrufo and Meson, 1996; Franck *et al.*, 1997). In other ecosystems, the most important effect of elevated CO_2 on decomposition may be indirect e.g., a rise in CO_2 might decrease the success of an abundant species which would alter ecosystem level litter chemistry. Elevated CO_2 can also lead to changes in the physical and biotic aspects of the environment and alter biomass allocation patterns. These changes have the potential to affect litter decomposition (Dukes and Hungate, 2002).

Elevated atmospheric CO_2 will not have any direct impact on the soil community, as the concentration of CO, within the soil atmosphere is sufficiently high to be affected by the predicted increase in atmospheric CO_2 (O'Neill, 1994). However, there are several indirect effects which may influence soil community that could potentially alter rates of carbon turnover in the terrestrial ecosystem (Sowerby *et al.*, 2000). Under elevated CO_2 consistently greater microbial biomass than ambient soil samples has been reported (Diaz *et al.*, 1993; Zak *et al.*, 1993 Schenk *et al.*, 1995. On the contrary, Kampichler *et al.*, (1998) and Runion *et al.* (1994) have not reported any increase in the soil microbial biomass in response to elevated CO_2 concentration.

Diaz *et al.* (1993) hypothesized that the extra carbon resulting from elevated levels of atmospheric CO_2 could be allocated to the microbial biomass in soils with carbon and nutrient accumulation in soil organic matter. According to this hypothesis, increased carbon inputs would stay within the soil system and not return to the atmosphere immediately (Sowerby *et al.*, 2000). So a greater resource base for supporting microbes would be produced and increased microbial biomass could result (van Ginkel and Gorissen, 1998).

Physiological responses of plants to rising CO_2 will affect belowground environment and may consequently affect decomposition as well (Tabak and Cooke, 1968; Dukes and Hungate, 2002). Carbon dioxide-driven increases in soil moisture (Bremer *et al.*, 1996; Niklaus et al., 1998) may also impact SOM decomposition (and possibly root litter decomposition) in several systems.

Thus, there are several mechanisms through which rise in atmospheric CO_2 could cause biologically important effects on decomposition. Plant physiological responses to rising CO_2 will affect belowground environment and may consequently affect decomposition (Dukes and Hungate, 2002). Changes in the availability of carbon and soil moisture could directly affect the activities of decomposers, or they could affect the composition and size of microbial community. In many systems, soil respiration and microbial respiration have been shown to increase in response to elevated CO_2 suggesting increase in carbon inputs to soil followed by its metabolization by microbial community (Zak *et al.*, 2000). Carbon dioxide — driven increases in soil moisture (Bremer *et al.*, 1996; Niklaus *et al.*, 1998) may also impact SOM decomposition (and possibly root litter decomposition) in several systems.

6.6. Soil texture and organic carbon

There are several important edaphic and physical factors that strongly influence the accumulation and decomposition of organic matter in soils (Agren *et al.*, 1996). The biological stability of soil organic carbon (SOC) is influenced by its chemical structure and the existence of various mechanisms of protection offered by soil minerals and their spatial arrangement within the soil matrix (Baldock and Skjemstad, 2000). In addition to defining the potential availability of SOC to decomposer organisms living in soils, the chemical structure of SOC also influences the strength with which mineral and organic soil components interact (Krull *et al.*, 2001). The degree of physical protection of SOC is mainly a function of soil texture, specific mineral surface area, and soil mineralogy. However, other soil parameters (e.g., water holding capacity, pH and porosity) can act as rate modifiers in attaining the protective capacity set by the mineral matrix of soil (Kroll *et al.*, 2001). The characteristics of soil matrix, such as clay, aluminium and iron contents and soil pH moderate soil carbon turnover rates in numerous ways (Dalal, 2001).

Clay minerals reduce the decomposition of soil organic matter by stabilizing part of organic matter through clay-organic matter complexes and cation bonding (Oades, 1995). Intercalation of organic matter in smectite clays and adsorption of carbohydrates on to clay surfaces also results in substantial soil organic matter stabilization (Oades, 1988; Anderson, 1992). Chapin *et al.* (2002) pointed out that clay alters the physical environment of soils by increasing water holding capacity and the resulting reduction in oxygen supply reduces the decomposition in wet clay soils. Even at moderate levels of soil moisture, clays enhance organic accumulation by binding soil organic matter (making it less accessible to microbial enzymes), binding microbial enzymes (reducing their effectiveness in breaking down substrates), and binding the soluble product of exoenzyme activity (making these products less available for absorption by soil microbes). This binding of organic matter to clay occurs due to attraction of the positive charges on the organic matter (amine groups) by the negatively charged sites on clay minerals. Clays can also form bridges with polyvalent cations (Ca^{2+}, Fe^{2+}, Al^{3+}, Mn^{4+}) that bind to negative groups (e.g., carboxyl groups) on organic matter (Stevenson, 1994). The net effect of this binding is protection of soil organic matter and reduction in decomposition rate (Baldock and Skjemstad, 2000). Soil organic matter protection by clay minerals is most important in grassland ecosystems or in tropical forests, where decomposition is relatively rapid and where soil animals rapidly mix fresh litter with mineral soil. Mineral protection of soil organic matter is less important in conifer forests or tundra where much of the decomposition occurs above the mineral soil in a well developed organic 0-horizon. Thus, it appears that both type and quality of clay influence the process of decomposition.

The architecture of the soil mineral matrix, i.e., arrangement of pores and soil particles influence the biological stability of organic materials through its effect on water and oxygen availability, entrapment and isolation of decomposers, and through the dynamics of soil aggregation (Krull *et al.*, 2001). Adequate quantities of available water and oxygen are required to optimise the processes of decomposition and mineralization. The pore size distribution of a soil also influences the ability of decomposer organisms to reach potential organic substrates (Krull *et al.*, 2001). Kilbertus (1980) suggested that bacteria can only enter pores > 3 mm. Decomposition of organic carbon within pore sizes less than this lower limit can only occur via diffusion of extracellular enzymes away from organisms towards a substrate and then diffusion of the products of enzyme reactions back to the organisms. With increasing clay content, proportion of the total porosity found in small pores increases, and the potential stabilization of organic carbon against biological attack due to the exclusion of decomposer organisms increases (Krull *et al.*, 2001). Protozoa and nematodes are excluded from pores less than 5 mm and 30 mm, respectively. Thus, organic carbon residing in pores smaller than these diameters in the form of molecules, small particles or bacterial or fungal tissues will not be susceptible to decomposition or predation by soil fauna (Krull *et al.*, 2001).

Organic carbon is not uniformly distributed in soil matrices instead exists as discrete particles, masses of amorphous materials typified by the mucilage exuded by microorganisms, or individual molecules adsorbed to mineral particles (Ladd *et al.*, 1993; Ransom *et al.*, 1997). Thus, encapsulation of organic carbon by flocculation of clay particles, adsorption of mineral particles around organic particles, or formation of stable aggregates will influence the biological stability of organic carbon as it places a physical barrier between potential substrates and decomposer organisms or their extracellular enzymes. Encapsulation can occur at size scales ranging from nanometres (e.g. encapsulation of organic carbon into pores between packets of clay particles) to centimetres (e.g. encapsulation of a piece of plant residue by mineral particles). Soil mineral particles are typically bound together into larger secondary particles referred to as aggregates (Krull *et al.*, 2001) which are affected by chemical, microbial, plant and physical processes (Lynch and Bragg, 1985; Glinski and Lipiec, 1990). Amelung and Zech (1996) demonstrated the influence of aggregation on the biological stability of organic carbon. They showed that organic carbon buried within aggregates was associated with organic materials having a higher C:N ratio, higher content of less biologically altered lignin, and a higher content of neutral sugars than organic carbon associated with the 0-0.5 mm external layer of aggregate surfaces. In addition soil organic matter stabilization is also influenced by non-biological reactions between N and organic matter (Mortland and Wolcott, 1965; Johnson, 1992; Agren *et al.*, 1996; Paul and Clark, 1996).

6.7. Disturbance

Pickett and White (1985) have defined disturbance as "... any relatively discrete event in time that disrupts ecosystem, community, or population structure and changes resources, substrate availability, or the physical environment." Natural and anthropogenic disturbances, such as fires, windstorms, landslides, flooding, logging, outbreaks of pathogens and herbivory, are all drastic events that in some way change ongoing ecosystem processes and force plants and animals to adjust to these new conditions (Grime, 1979). Since plant communities are not as mobile as animals, they are often more directly affected. Although disturbances often operate in long time intervals (up to 200 years) and often not in a regular periodicity, the impact is of such magnitude that it shapes the plant community for a long time afterwards. This makes disturbance a key ecological factor in various communities and ecosystems. Studies to document effect of disturbances on decomposition in different ecosystems, recently conducted by Neher *et al.*, (2003) in north Carolina, involved comparative decomposition of museum board (predominantly cellulose) and balsa wood substrates in 18 sites chosen to represent two disturbance levels nested within three ecosystems: agriculture, wetland and forest. More rapid decomposition of cellulosic substrates occurred in disturbed than undisturbed sites (Table 6.1), an observation also reported by Visser *et al.* (1983).

Percentage of the substrate remaining (museum board or balsa wood) was correlated negatively with pH for all sites, suggesting that pH should be included as a covariate if measures of decomposition are used as environmental indicators. The study enabled to conclude that measures of decomposition of predominantly cellulose substrates can be used to distinguish between relative levels of disturbance in agricultural and wetland but not forest systems. Differences in decomposition may signal either a change in decomposer community

Table6.1. Relationship between percentage mass remaining of museum board and balsa wood substrates vs. days of incubation in the field in relatively disturbed and undisturbed sites in north Carolina.

Substrate	**Ecosystem**	**Condition**	**Intercept**	**Slope**
Museum board	Agriculture	Disturbed	55.84"	-0.092"
		Undisturbed	89.47	-0.278
	Forest	Disturbed	86.77	-0.147
		Undisturbed	90.43	-0.124
	Wetland	Disturbed	47.55*	-0.147
		Undisturbed	88.15	-0.238
Balsa wood	Agriculture	Disturbed	93.26	-0.109
		Undisturbed	94.42	-0.128
	Forest	Disturbed	96.54	-0.089
		Undisturbed	100.24	-0.082
	Wetland	Disturbed	92.87	-0.183
		Undisturbed	101.43	-0.132

* Significant at $P \leq 0.05$ between disturbed and undisturbed sites.

Source: Neher *et al.* (2003)

or condition of biotic and abiotic resources at a site (Neher et al., 2003). Mechanisms by which disturbance stimulates decomposition is basically same at all scales, ranging from the movement of earthworms through soils to tillage of agricultural fields. Disturbance disrupts soil aggregates so the organic matter contained within them becomes more exposed to oxygen and microbial colonization. This disturbance effect is most pronounced in warm wet soils, where increased aeration has greatest effect on decomposition. In addition, soil disturbance increases decomposition by promoting aeration and exposing new surfaces to microbial attack.

•••

CHAPTER 7

HUMUS BIOSYNTHESIS

7.1. Nature of humic substances

Humus, or natural soil organic matter as it is alternatively called, is the most important natural resource of fundamental importance to soil fertility (Tan, 2003; Wershaw, 2004; Peña-Méndez *et al.*, 2005). Theoretically, humus can contain any of the organic compounds synthesised by plants or animals. By the process of mineralization, part of organic carbon from the plant (and animal) residues is converted into carbon dioxide, part of it is metabolised by soil (micro-) organisms and thereafter converted into their biomass and part is transformed through the processes of biodegradation, re-polymerisation and reactions with soil inorganic constituents to humus, or soil organic matter.

Types and properties

Humic substances are the most widely distributed products of biosynthesis on the Earth (Tan, 2003; Peña-Mendez *et al.*, 2005). They have been shown to contain a wide variety of molecular components and some of them are: polysaccharides, fatty acids, polypeptides, lignins, esters, phenols, ethers, carbonyls, quinones, lipids, peroxides, various combinations of benzene, acetal, ketal, and furan ringed compounds and aliphatic (carbon chains) compounds. The oxidative degradation of some humic substances produces aliphatic, phenolic, and benzene carboxylic acids in addition to n-alkanes and n-fatty acids. The major phenolic acids released contain approximately 3 hydroxyl (-OH) groups and between 1 and 5 carboxyl (-COOH) groups. Humic substances are sub-divided into three major fractions: (i) humic acids (ii) fulvic acids, and (iii) humin. These sub-divisions are arbitrarily based on the solubility of each fraction in water adjusted to different acid-alkaline (pH levels) conditions (Ghabbour and Davies, 2001).

Humic acids

Humic acids comprise a mixture of weak aliphatic (carbon chains) and aromatic (carbon rings) organic acids which are not soluble in water under acid conditions but are soluble in water under alkaline conditions. Humic acids are precipitated from aqueous solution when the pH is decreased below 2. They are termed polydisperse because of their variable chemical features. From a three dimensional aspect these complex carbon containing compounds are considered to be flexible linear polymers that exist as random coils with cross-linked bonds. On an average 35% of the humic acid molecules are aromatic (carbon rings), while the remaining compounds are in the form of aliphatic (carbon chains) molecules. Thus, they are thought to be complex aromatic macromolecules with amino acids, amino sugars, peptides and aliphatic compounds involved in linkages between the aromatic groups. The molecular size of humic acids range from approximately 10,000 to 100,000. flumic acid polymers readily bind clay minerals to form stable organic-clay complexes and also form salts with inorganic trace mineral elements. These trace elements are bound in humic acid molecules in a form that can be readily utilized by various living organisms. As a result humic acids function as important ion-exchange and metal-complexing (chelating) systems. Due to the development of analytical techniques, great efforts have been made to elucidate the molecular structures of humic acids. From the work of Stevenson (1994), to more recent models of Schulten (2000, 2001), Kujawinski *et al.* (2002a, 2002b) and Stenson *et al.* (2002, 2003), several molecular structures describing the structure of humic acids have been proposed. The hypothetical structure for humic acid (Fig. 7.1) contains free and bound phenolic OH groups, quinone structures, nitrogen and oxygen as bridge units and COOH groups variously placed on aromatic rings.

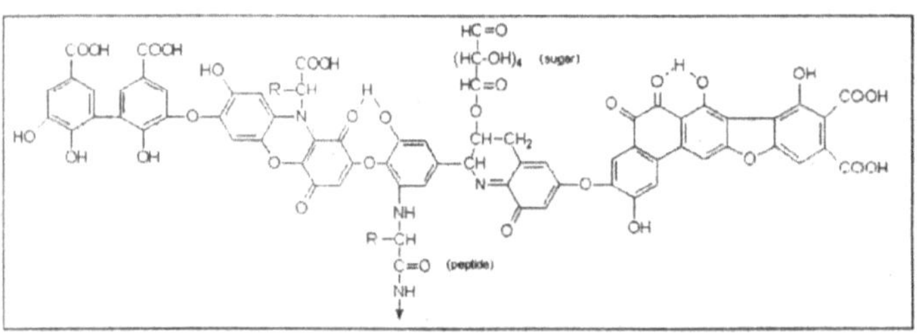

Fig. 7.1. Model structure of humic acids as proposed by Stevenson (1994).

Fulvic acids

Fulvic acids are a mixture of weak aliphatic and aromatic organic acids which are soluble in water at all pH conditions (acidic, neutral and alkaline). Their composition and shape is quite variable. The size of fulvic acids is smaller than humic acids and molecular weights range from approximately 1,000 to 10,000. Fulvic acids have oxygen content twice that of humic acids. They have many carboxyl (-COOH) and hydroxyl (-COH) groups which make them chemically much more the reactive. The exchange capacity of fulvic acids is more than double humic acids due to the total number of carboxyl (-COOH) groups present which in fulvic acids range from 520 to 1120 cmol (H+)/kg. Fulvic acids collected from many different sources and analysed show no evidence of methoxy (-CH3) groups. In addition, they are low in phenols, and are less aromatic compared to humic acids.

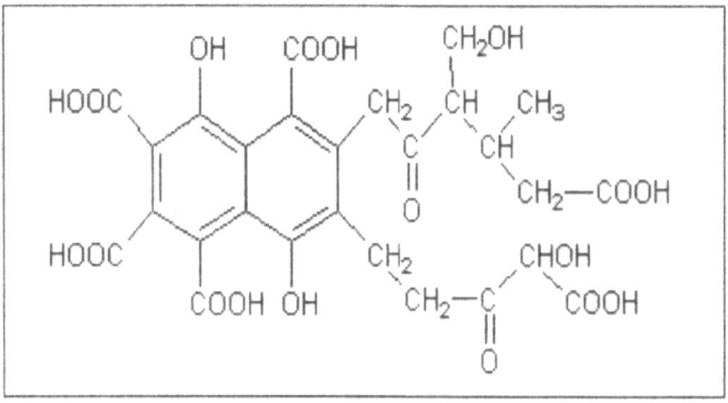

Fig. 7.2. Model structure of fulvic acids.

Because of their relatively small size, fulvic acid molecules can readily enter plant roots, stems, and leaves and as they enter these plant parts they carry trace minerals from plant surfaces into plant tissues. Fulvic acids are key ingredients in high quality foliar fertilizers. The hypothetical model structure of fulvic acid (Buffle's model) contains both aromatic and aliphatic structures, both extensively substituted with oxygen-containing functional groups (Fig. 7.2).

Humin

Humin is that fraction of humic substances which is neither soluble in alkali (high pH) nor in acid (low pH). Humin complexes are considered macro-organic substances because their molecular weights range from approximately 100,000 to 10,000,000. The chemical and physical properties of this fraction are only partially understood largely because of the extraction problems. Of all the humic substances, humin is most resistant to decomposition in soil.

Many investigators now believe that all dark coloured humic substances are part of a system of closely related, but not completely identical, high molecular weight polymers. According to this concept, differences between humic acids and fulvic acids can be explained by variations in molecular weight, numbers of functional groups (carboxyl, phenolic OH) and extent of polymerisation as depicted in Fig. 7.3.

Carbon and oxygen contents, acidity and degree of polymerisation, all change systematically with increasing molecular weight. The low molecular weight fulvic acids have higher oxygen but lower carbon contents than the high molecular weight humic acids. Fulvic acids contain more functional groups of an acidic nature, particularly COOH. The total acidities of fulvic acids (900 - 1400 meq/100g) are considerably higher than humic acids (400 - 870 meq/100g). Another important difference is that while the oxygen in fulvic acids can be accounted for largely in known functional groups (COOH, OH, C=O), a high proportion of the oxygen in humic acids seems to occur as a structural component of the nucleus (Stevenson, 1994).

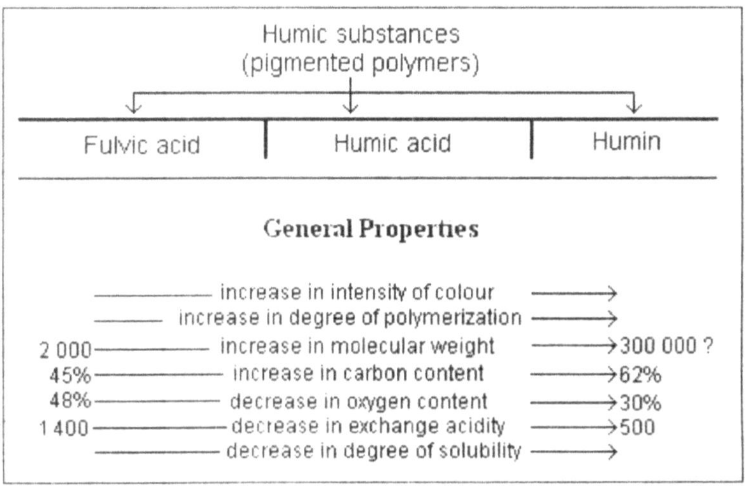

Fig. 7.3. Summary of the major features of humic substances.

7.2. Synthesis of humic substances

Synthesis of humic substances is one of the least understood and most intriguing aspects of humus chemistry. Early workers (for example, Kononova, 1966) understood that humus arises mainly from the degradation of the dead plant tissues with minor contributions from decaying animal remains as well. Many of these early workers suggested that humus was composed of the end products of synthetic reactions that alter the structures of plant degradation products. In contrast, a number of workers recently have proposed that humic substances consist mainly of the partial degradation products of plant polymers (Wershaw, 1986; Wershaw, 1994; Piccolo, 2001). This controversy has led to the development of two different types of models of humus biosynthesis: (i) the humic polymer model (Stevenson, 1994), and (ii) the molecular aggregate model (Wershaw, 2004).

Humic polymer model

In this model it is postulated that the component molecules of humus are polymeric species that have been synthesized by secondary synthesis reactions from plant degradation products. Stevenson (1994) provided a detailed exposition of the synthetic pathways that lead to the formation of humic substances and the main pathways are shown in the Fig. 7.4.

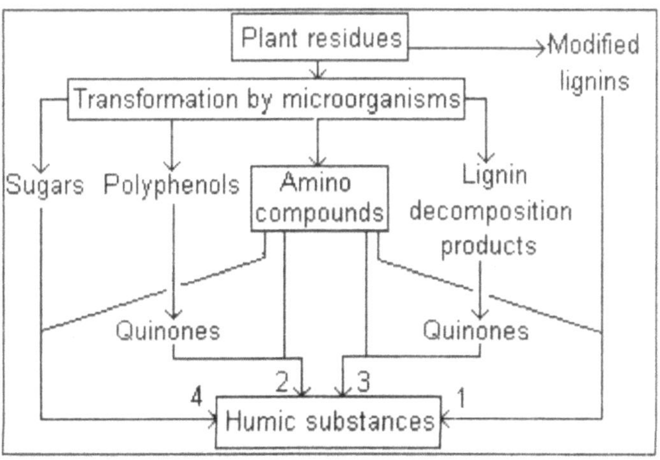

Fig. 7.4. Pathways of humus biosynthesis.

The classical theory, popularised by Waksman, is that humic substances represent modified lignins (pathway 1) but majority of the present-day investigators favour a mechanism involving quinones (pathways 2 and 3). In practice all four pathways must be considered as likely mechanisms for the synthesis of humic and fulvic acids in nature, including sugar-amine condensation (pathway 4). The four pathways may operate in all soils, but not to the same extent or in the same order of importance. A lignin pathway may predominate in poorly drained soils and wet sediments (swamps, etc.) whereas synthesis from polyphenols may be of considerable importance in certain forest soils. The frequent and sharp fluctuations in temperature, moisture and irradiation in terrestrial surface soils under a harsh continental climate may favour humus synthesis by sugar-amine condensation.

Lignin theory

For many years it was thought that humic substances were derived from lignin (pathway 1). According to this theory, lignin is incompletely utilized by microorganisms and the residum becomes part of the soil humus. Modifications in lignin include loss of methoxyl (OCH_3) groups with the generation of o-hydroxyphenols and oxidation of aliphatic side chains to form COOH groups. The modified material is subject to further unknown changes to yield first humic acids and then fulvic acids. This pathway, illustrated in the Fig. 7.5, is exemplified by Waksman's lignin-protein theory. Although lignin is less easily attacked by microorganisms than other plant components, mechanisms exist in nature for its complete aerobic decomposition. Otherwise un-decomposed plant remains would accumulate on the soil surface and the organic matter content of the soil would gradually increase until CO_2 was depleted from the atmosphere. The ability of soil organisms to degrade lignin has been underestimated in some quarters and its contribution to humus has been exaggerated. In normally aerobic soils lignin may be broken down into low-molecular-weight products prior to humus synthesis. On the other hand, the fungi that degrade lignin are not normally found in excessively wet sediments. Accordingly, it seems logical to assume that modified lignins may make a major contribution to the humus of peat, lake sediments, and poorly drained soils.

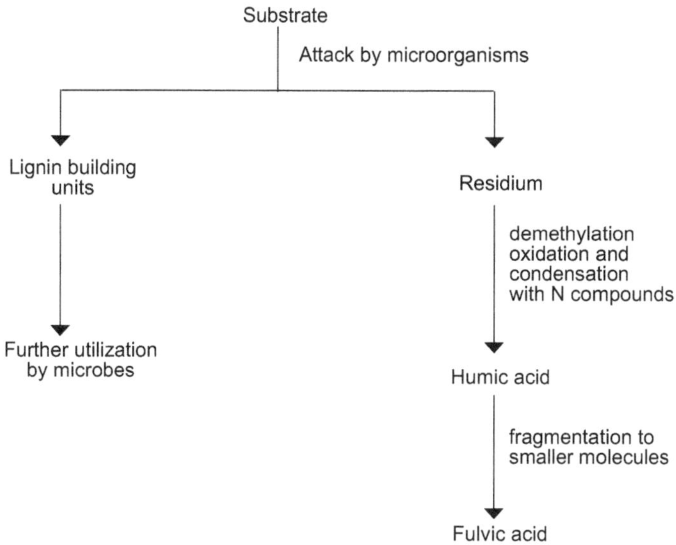

Fig. 7.5. Diagrammatic representation of lignin theory of humus biosynthesis

Polyphenol theory

In pathway 3 (Fig.7.4) lignin still plays an important role in humus synthesis, but in a different way. In this case phenolic aldehydes and

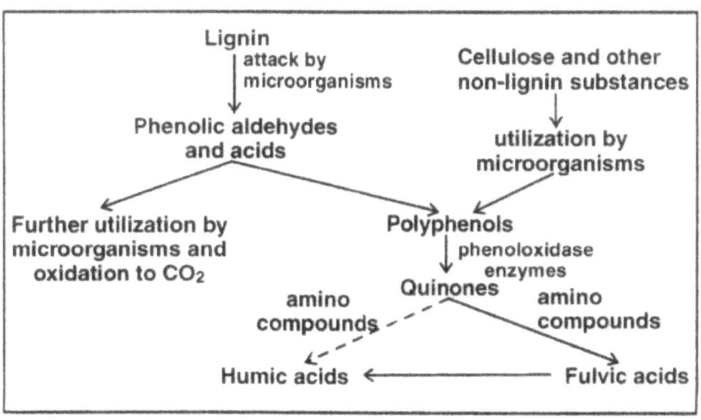

Fig. 7.6. Diagrammatic representation of polyphenol theory of humus biosynthesis.

acids released from lignin during microbial attack undergo enzymatic conversion to quinones, which polymerise in the presence or absence of amino compounds to form humic like macromolecules.

Pathway 2 (Fig. 7.4) is somewhat similar to pathway 3 except that the polyphenols are synthesized by microorganisms from non-lignin C sources (e.g., cellulose). The polyphenols are then enzymatically oxidized to quinones and converted to humic substances (Fig. 7.6). As noted earlier, the classical theory of Waksman is now considered obsolete by many investigators and according to current concepts quinones of lignin origin, together with those synthesized by microorganisms, are the major building blocks from which humic substances are formed. Possible sources of phenols for humus synthesis include lignin, microorganisms, uncombined phenols in plants and tannins. Of these, only the first two have received serious attention.

Sugar-amine condensation theory

The notion that humus is formed from sugars (pathway 4) dates back to the early days of humus chemistry. According to this concept reducing sugars and amino acids, formed as by-products of microbial metabolism, undergo non-enzymatic polymerisation to form brown nitrogenous polymers of the type produced during dehydration of certain food products at moderate temperatures. A major objection to this theory is that the reaction proceeds rather slowly at the temperatures found under normal soil conditions. However, drastic and frequent changes in the soil environment (freezing and thawing, wetting and drying), together with the intermixing of reactants with mineral material having catalytic properties, may facilitate condensation. An attractive feature of the theory is that the reactants (sugars, amino acids etc.) are produced in abundance through the activities of microorganisms.

The initial reaction in sugar-amine condensation involves addition of the amine to the aldehyde group of the sugar to form the n-substituted glycosylamine. The glycosylamine subsequently forms the N-substituted-1-amino-deoxy-2-ketose and the latter upon fragmentation yields 3-carbon chain aldehydes and ketones, such as acetol, diacetyl etc. and upon dehydration yields reductones and hydroxymethyl furfurals. All these compounds are highly reactive and readily polymerise in the presence of amino compounds to form brown-coloured products (Fig. 7.7).

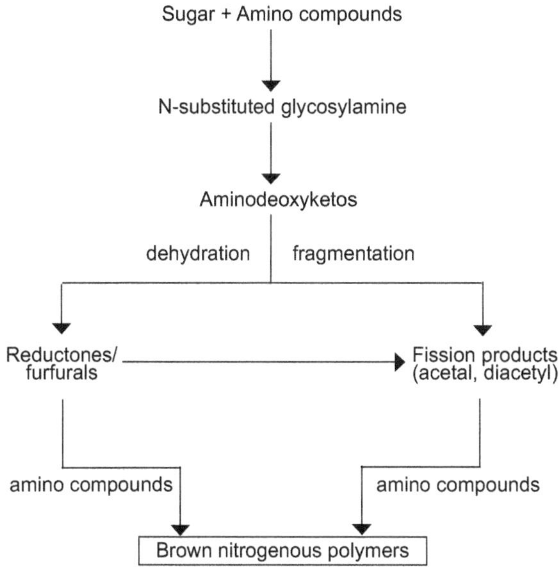

Fig.7.7. Diagrammatic representation of sugar-amine condensation theory of humus biosynthesis.

Molecular aggregate model

In this model, humus is depicted as being composed of molecular aggregates (supramolecular aggregates) of plant degradation products held together by non-covalent bonds (Wershaw, 1994) and humification is recognised as a three-step process of degradation, aggregation of the degradation products and degradation of the aggregation products. In solution, supramolecular aggregation is a reversible process in which the molecules are constantly combining and breaking apart (Lou *et al.*, 2004). Degradation pathways of the major components of plant tissues are discussed below along with the likely mechanisms of interactions of the resulting degradation products.

Degradation of plant residues

In natural systems the components comprising the plant tissues can undergo three possible types of degradation reactions: (i) biotic (enzymatically catalysed) reactions, (ii) pyrolytic reactions, and (3) abiotic reactions other than pyrolytic reactions. However, it appears that the degradation of plant tissues is mainly the result of enzymatically catalysed reactions (Wershaw, 2004).

Biodegradation of cellulose, the main structural component of vascular plants, has been reviewed by Perez *et al.*, (2002). Eubacteria and fungi degrade it to carbon dioxide and water under aerobic conditions while as protozoa and slime moulds degrade cellulose to methane and water under anaerobic conditions. These organisms bring about the cellulose degradation through the production of hydrolytic enzymes. For example, cellulases hydrolyse β-1-4-glycosidic bonds and two types of cellulases have been identified: endoglucanases and cellobiohydrolases (exoglucanases). Former hydrolyse internal cellulose bonds while the latter attack the ends of the cellulose chains. Both these types of enzymes degrade the cellulose chains to the disaccharide cellobiose. In order to use hydrolysis products as energy source, microorganisms should be able to cleave cellobiose into two glucose molecules by β-glucosidase.Anaerobic degradation of cellulose is less well studied. However, a strict-anaerobe *Clostridium thermocellum* is reported to bring about enzymatic degradation of cellulose. The enzymes in this species are packaged in cellulosomes that are attached to the cell surfaces of the organisms. Furthermore, the cellulase enzymes are associated with the protein scafoldin that localizes the various enzymes so that they act more effectively in concert (Atalla, 1999). Hemicelluloses, found in intimate contact with celluloses in higher plants, are degraded by hemicellulases to monomeric saccharides and acetic acid (Perez *et al.*, 2002). Studies have revealed that four different enzymes namely, endo-1-4-β-xylanase, acetyl esterase, α-glucuronidase and β-xylosidase are required to degrade O-acety1-4-0-methylglucuronxylan, the most common hemicellulose. Microbial degradation of pectic polysaccharides is also catalysed by a group of specialized enzymes called as pectinases (Beldman *et al.*, 1996; Benen *et al.*, 1996; Hoondal *et al.*, 2002). Like cellulose and hemicellulose, biological degradation ultimately leads to depolymerization of pectins to simple sugars. Apart from microbial degradation, pyrolysis of cellulose has been studied more extensively than that of any other biomass constituent. Shafizadeh (1982, 1984) has identified two different pathways for cellulose pyrolysis. First, at temperatures below about

300°C bond scission, formation of free radicals, dehydration, and formation of carbonyl, carboxyl, and hydroperoxide groups predominate followed by production of carbon monoxide, carbon dioxide, and char. Second pathway (Fig. 7.8) leads to formation of anhydro sugars, such as levoglucosan by dehydration (Nimlos et al., 2003)

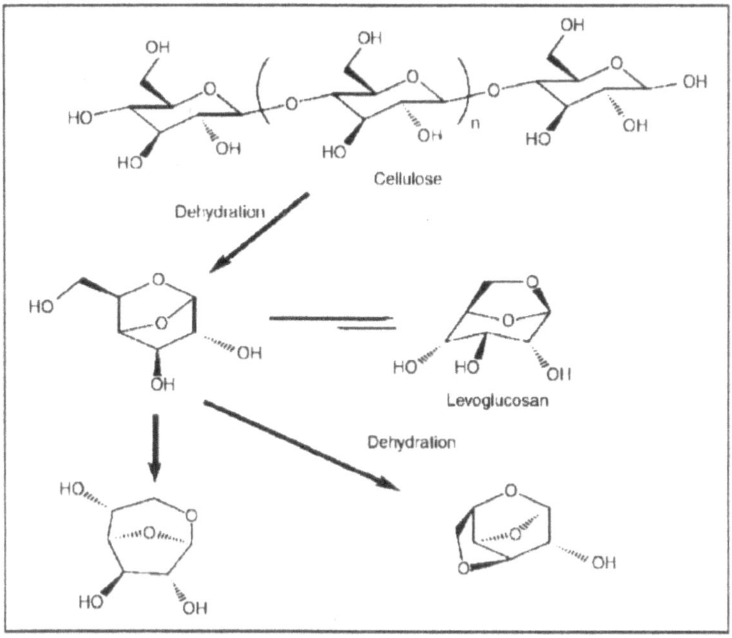

Fig.7.8. Pyrolysis of cellulose leading to formation of anhydro sugars.

Lignin, another structural constituent of vascular plants, is composed of varying amounts of *p*-hydroxyphenylpropanoid, guiacylpropanoid and syringylpropanoid monomeric units (Douglas, 1996). Although β-0-4 ether linkages are common in lignin (Fig. 7.9) but other linkages such as α-O-4, β-β, β-5, and 5-5 are also present (Saake *et al.*, 1996).

Since lignin occurs complexed with hemicellulose, lignin degradation cannot be considered independently of the hemicellulose degradation. Therefore, degradation of lignin requires prior degradation of hemicellulose. Degradation of lignin brought about by microorganisms through extra-cellular enzymes generally involves both depolymerization

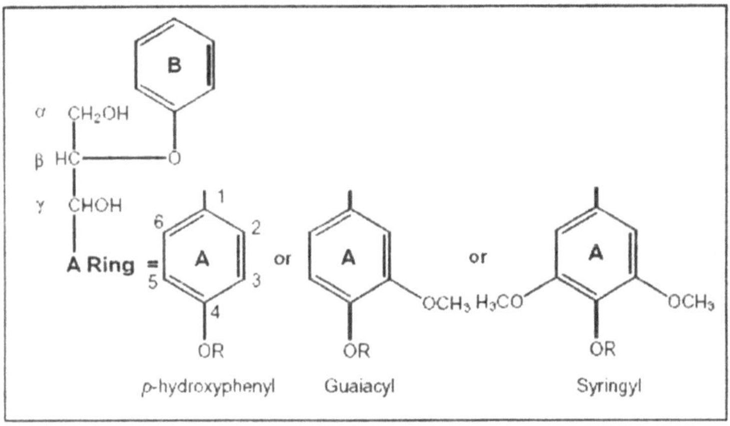

Fig.7.9. Lignin monomeric units.

and aromatic-ring cleavage (Crestini *et al.*, 1998). Using lignin model compounds, Crestini *et al.* (1996) pointed that the white-rot fungus *Lentinus edodes* brings about cleavage of aromatic rings coupled with β-O-4 oxidation. The enzymatic aromatic ring cleavage mostly follow the β-ketoadipate pathway (Fig. 7.10) and can take place either between two hydroxyl groups (*ortho*-cleavage) or adjacent to a hydroxyl group (*meta-cleavage*). Schemes I and II (*ortho*-cleavage) lead to the production of β-ketoadipate. Scheme III is the first step of metacleavage, and Scheme IV (Fig. 7.10) is the first step in the gentisate pathway in which fission takes place between the carbon atom that is attached to a carboxylate group and an adjacent phenolic carbon. Schemes III and IV do not lead to the formation of β-ketoadipate (Mohamed *et al.*, 2001 and Zaar *et al.*, 2001).

Likewise other chemical constituents of the detritus like cutin, subrin (Kolattukudy, 2001), aliphatic acids (Ohta, 1989), plant waxes (Kolattukudy and Espelie, 1989) and secondary metabolite's like tannins (Gross, 1999) and terpenes (Claeys *et al.*, 2004) also contribute various degradation products that aid in the formation of humus.

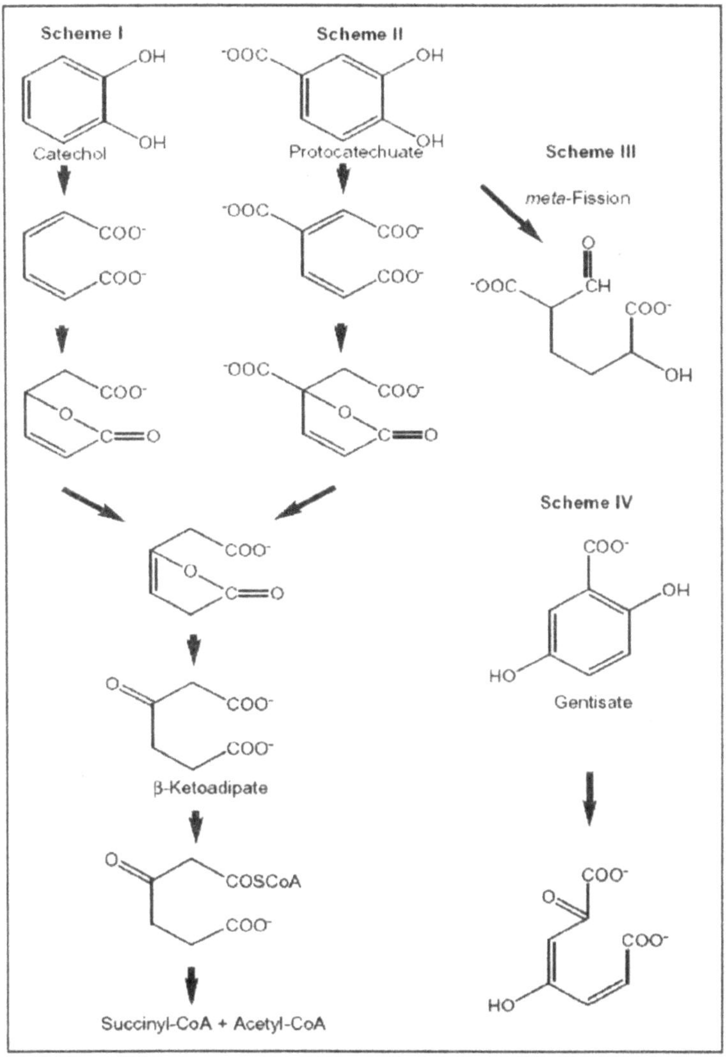

Fig.7.10. Pathways of lignin degradation.

Interaction of degradation products

Molecular aggregate model posits that monomeric or oligomeric products of biodegradation of the major components of plant tissues mix to form new supramolecular aggregates. The process of building these new supramolecular aggregates results in the formation of humus. In general, with the exception of polyphenolic compounds (tannins), non-covalent interactions of degradation products are stated to occur during the process of humification (Wershaw, 1994; Piccolo, 2001). In fact, Wershaw (1994) proposed that depolymerization and oxidation reactions that take place during enzymatic degradation of biopolymers produce amphiphile-molecules that have a polar (hydrophilic) part and a nonpolar (hydrophobic) part. These amphiphiles spontaneously assemble into ordered aggregates in which the hydrophobic parts of the molecules form the interiors and the hydrophilic parts of the molecules make up the exterior surfaces of the aggregates. These ordered aggregates constitute the humus in soils and sediments. Piccolo (2001) also proposed that supramolecular associations of self-assembling heterogeneous and relatively small molecules derived from the degradation and decomposition of dead biological material leads to formation of humus with hydrophobic dispersive forces holding together the supramolecular aggregates. Studies of Haslam (1998) and Wershaw and Kennedy (1998) have revealed that hydrolysable and nonhydrolysable tannins are of particular importance in understanding the aggregation of humus because they form strong non-covalent bonds with a variety of different types of compounds and, in particular, with carbohydrates and proteins.

7.3. Forms of humus

Humus forms can be defined as morphological patterns observed in the association of organic matter with mineral matter at the top of soil profiles (Ponge, 2003). Most of the workers distinguish between mull, moder and mor humus types. These humus forms are the seat of most biological transformations taking place in terrestrial ecosystems, being at the interface between plants, animals and microbes. Main biological characteristics of the three main forms of humus are given in the Table 7.1.

Table 7.1. Main characteristics of mull, moder and mor humus forms.

	Mull	**Moder**	**Mor**
Ecosystem	grasslands, deciduous woodlands with rich herb layer, Mediterranean scrublands	deciduous and coniferous woodlands with poor herb layer	heathlands, coniferous woodlands, sphagnum bogs, alpine meadows
Biodiversity	high	medium	low
Productivity	high	medium	low
Litter horizon	OL,OF	OL,OF,OH	OL,OM
Soil type	brown soils	grey-brown podzolic soils	podzols
Phenolic content of litter	poor	medium	high
Humification	rapid	slow	very slow
Humified organic matter	organo-mineral aggregates with clay-humus complexes	holorganic faecal pellets	slow oxidation of plant debris
Exchange sites	mineral	organic (rich)	organic (poor)
Mineral weath-ering	high	medium	poor
Mineral buffer type	carbonate range	silicate range	iron/aluminium range
Impact of fire	low (except in Mediterranean ecosystems	medium	high
Regeneration of trees	easy (permanent)	poor (cyclic processes	none (fire needed)
Dominant mycorrhizal types	arbuscular mycorrhizas	ectomycorrhi-zas	ericoid and arbutoid mycorrhizas
Mycorrhizal patterns	Zygomycetes	Basidiomycetes	Ascomycetes

Contd...

Nitrogen forms	protein, ammonium and nitrate	protein, ammonium	protein
Nutrient availability to plants	direct	indirect (through extramatrical mycelium)	poor
Fauna	low	macrofauna (poor), mesofauna (rich), microfauna	mesofauna (poor), microfauna (poor)
Faunal group dominant in biomass	mega-, macro-, meso-, micro-fauna	enchytraeids	none
Microbial group dominant in biomass	earthworms	fungi	none
Affinities with polluted condition	low	medium	high

Source: Ponge (2003)

7.4. Role of humic substances

Soil fertility

Humic substances play an important role from the agronomical point of view (Peña-Mendéz *et al.*, 2005) and are a good source of energy for soil organisms. Humic substances and non-humic (organic) compounds provide energy and many of the mineral requirements for soil microorganisms and soil animals. Soil organisms that lack the photosynthetic apparatus to capture energy from the sun survive on residual carbon containing substances on or in the soil. These organisms, in turn, perfonn many beneficial functions which influence soil fertility and plant health (Clapp *et al.*, 2001; Huang, 2002). For example, the bacteria release organic acids which aid in the solubilisation of bound mineral elements in soil and also release complex polysaccharides that help create soil crumbs (aggregates) which give soil a desirable structure. Other beneficial soil microorganisms, such as the actinomycetes release antibiotics into the soil. These antibiotics are taken up by the plants to protect themselves against pests. In addition, antibiotics function to create desirable ecological balances of soil organisms on the root surface (rhizoplane) and in soil near the root (rhizosphere). Fungi also perform many beneficial functions in soils. For example, mycorrhizas aid plant roots in the uptake of essential mineral nutrients and water. Other fungi decompose plant residues releasing bound nutrients for other organisms. Some soil animals create tunnel-like channels in the soil which aid in movement of water and gas exchange. Thus, a healthy fertile soil must contain sufficient carbon containing compounds to sustain the billions of microscopic life forms required for a healthy plant.

Humus functions to improve the water holding capacity of soils through its influence on the development of a desirable soil structure that facilitates water infiltration and helps hold water within the root zone (Piccolo and Mbagwu, 1999). Because of their large surface area and internal electrical charges, humic substances function as water sponges and have the ability to hold water seven times their volume, a greater water holding capacity than soil clays. This is why growers who apply humate-based fertilizers and integrate production practices which preserve humic substances, can frequently harvest a crop during periods of dry weather. Complex carbohydrates synthesized by bacteria and humic substances function together with clay and silt to form soil aggregates. As the humic substances become intimately associated with the mineral fraction of the soil, colloidal complexes of humus-clay and humus-silt aggregates are formed. These aggregates are formed by electrical processes which increase the cohesive forces that cause very fine soil particles and clay components to attract each other. Once formed these aggregates help create a desirable crumb structure in the top soil, making it more friable. Soils with good crumb structure have improved tilth, and more porous openings (open spaces). These pores allow for gaseous interchange with the atmosphere, and for greater water infiltration. Mean residence time of these organo-mineral complex aggregates varies and based on radiocarbon dating, using extracts from non-disturbed soil, average residence time of humic substances is: humin 1140 years, humic acid 1235 years, and fulvic acid 870 years.

Degradation or inactivation of toxic substances, such as nicotine, aflatoxin, antibiotics, phenols, most pesticides, and also metals including radionuclides is also mediated by humic substances (Lubal *et al.*, 1998; Pacheco and Havel, 2001). In the microbial degradation process not all of the carbon contained within these toxins is released as CO_2 . A portion of these toxic molecules, primarily the aromatic ring compounds are stabilized and integrated within the complex polymers of humic substances. Humic substances have electrically charged sites on their surfaces which function to attract and inactivate pesticides and other toxic substances. For this reason the use of humates for clean up of toxic waste sites is recommended.

Soil enzymes are also stabilized and/or inactivated by humic substances. Stabilization of soil enzymes by covalent bonding renders these enzymes less vulnerable to microbial degradation. Once stabilized and bound to the humic substances, enzyme activity is greatly reduced or ceases to function (Sinsabaugh *et al.*, 2002). However, many of these bonds are relatively weak and during periods of pH change within the soil, these enzymes can be released. Humic substances also help to stabilize soil temperatures and slow the rate of water evaporation. Their insulating properties help maintain a more uniform soil temperature, especially during periods of rapid climate changes, such as cold spell or heat waves. Because of electrical charge, humic acids and fulvic acids are known to be involved in three specific chemical reactions: (a) electrostatic (columbic) attraction, (b) complex formation or chelation, and (c) water bridging.

Electrostatic attraction of trace minerals reduces their leaching into subsoil. The metal cations are loosely attached, thus can be released when attracted to another stronger electrical charge. Such cations are thus readily available in the soil environment for transport into the plant roots or exchange for other metal cations. A complex reaction with metal cations occurs on the surface of humic substance is termed as chelation. Two negatively charged sites on the humic substance attract metal cations with two positive charges. As a result the cation binds itself tc more than one charged anionic site. By forming organo-metal chelates, these organic acids bring about the dissolution of primary and secondary minerals within the soil (Tan, 1986; Chen and Aviad, 1990; Barker *et al.*, 1997). These minerals then become available for uptake by plant roots. The greater the affinity of the metal cation for humic acids or fulvic acids, the easier the dissolution of the cation from various mineral surfaces. Both the acidic and the chelation effects appear to be involved in dissolution of minerals and binding processes. Evidence for the dissolution of minerals can be supported by x-ray diffraction and infrared analysis. Chelation of plant nutrients, such as iron (Fe), copper (Cu), zinc (Zn), magnesium (Mg), manganese (Mn), and calcium (Ca) reduces their toxicity, prevents their leaching, and increases their uptake rate by plant roots.

The chelation exchange reaction involves a transition element and the release of these trace minerals into the plant is quite different from the classical cation exchange system. The cation with a plus two charge, present in the chelate cannot be replaced by a singly charged cation, such as H^+,K^+ or Na^+ Cations with one positive charge are unable to replace a metal ion, such as Cu^{++}, with two positive charges. The chelated metal ion can be exchanged by another transitional ion that has two positive charges. The chelates provide the carrier mechanism by which depleted nutrient elements are replenished at the root surface. This process also increases the mass flow of micro-nutrient mineral elements to the root. The chelation of heavy toxic metallic elements present within the soil is also influenced by the presence of humic substances. When toxic heavy metals, such as mercury (Hg), lead (Pb), and cadmium (Cd) are chelated, these organo-metal complexes become less available for plant uptake. Detailed studies of chelation of heavy metals in industrial sludge have illustrated the value of humic substances in preventing uptake of these toxic metals. Water bridging is an important function of humic and fulvic acids. It involves the attraction of a water molecule followed by the attraction of a mineral element cation (illustrated by ($^-COO—H_2O—Fe^+$) at an anionic site on the humic or fulvic acid polymers. The water holding capacity of humic substances and their ability to bind trace mineral elements function together in water bridging. It is believed to improve the mobility of nutrient ions through the soil solution to the root and also help reduce leaching of plant nutrients into the subsoil. Recent experiments indicate that water bridging may be more common in humic substances than originally believed.

Stored energy and trace mineral content of humic substances help sustain soil organisms involved in transmutation. The presence of humic substances within saline soils (those soils which contain high salt concentrations, e.g. sodium chloride) aid in the transmutation of the sodium ions. The transmutation reaction, a biological process that occurs within living organisms, results in combining the sodium with a second element, such as oxygen, to form a new element. Although the theory of transmutation has met considerable opposition by some traditional physicists and chemists, biologists have recorded convincing data to prove that transmutation occurs in living organisms. Application of humins, humic acids, and fulvic acids to saline soils, in combination with specific soil organisms, reduces the concentration of sodium salts (e.g. NaCI). By reducing the salt content of a soil, its fertility and health can be "brought back" to provide a more desirable environment for plant root growth.

Plant growth

Plant growth is influenced directly and indirectly by humic substances. Positive correlations between the humus content of the soil and plant yields and product quality have been established through many studies (Chen *et al.*, 1999). Indirect effects, as discussed above, are mediated by providing energy for the beneficial organisms within the soil, by influencing the water holding capacity, structure, release of plant nutrients from soil minerals, besides increase in the availability of trace elements leading to improvement in soil fertility. Direct effects include those changes in plant metabolism that occur following the uptake of macromolecules, such as humic acids and fulvic acids. One stimulative effect of humic substances on plant growth is enhanced uptake of major plant nutrients, such as nitrogen (N), phosphorus (P), and potassium (K) (Kapulnik, 1996; Day *et al.*, 2000;). In fact, requirement for N-P-K fertilizer applications is reduced when adequate humic substances are present within the soil. Some studies have also reported increased uptake of calcium (Ca), and magnesium (Mg) upon irrigation of plants with liquid suspensions of humic acids or fulvic acids. Another key benefit of humic substances, which maximizes fertilizer efficiency, is a reduction in the toxicity and leaching of nitrogen compounds into subsoil water. Humic substances hold these major plant nutrients in a molecular form which reduces their solubility in water. These binding processes also help prevent volatilisation into the atmosphere.

Humic substances have been shown to favourably influence seed germination and seedling development. As the humic substances enter the seed cells, respiration rate increases, and cell division processes are also accelerated. These same respiratory processes enhance root meristem development and activate other growing points within the seedlings. Humic substances have been demonstrated to enhance mitotic activity during cell division under carefully controlled experiments. However, excessive concentrations of humic and fulvic acids can inhibit seed germination and at high concentrations can even kill young seedlings.

Humic substances have a very pronounced influence on the growth of plant roots (Chen and Aviad, 1990; Chen *et al.*, 2001). When humic acids and/or fulvic acids are applied to soil, enhancement of root initiation and increased root growth are observed. In most experimental studies plant root growth is stimulated to a greater extent compared to stimulation of aboveground plant parts. Replicate treatments of plants grown within the greenhouse, with and without humic acids and fulvic acids revealed that treated root weights averaged from 20 to 50% heavier compared to the weights of non-treated roots. The type of humic substances applied had a significant influence on the extent of increase. Not all humic substances contain a desirable molecular mixture of humins, humic acids and fulvic acids capable of rapid stimulation of root growth. Some humic substances, because of their large molecular sizes fail to stimulate plant root development. Root stimulation occurs when the smaller molecular components within fulvic acids occur at a concentration ranging from 10 to 100 mg/litre of solution. Growth is further stimulated when fulvic acids are used in combination with humic acids and other required plant nutrients. It needs to be emphasized that excessively high concentrations of humic substances can result in a reduction in root weight. Thus, for optimum plant growth humic acids and fulvic acids should be applied at relatively low concentrations.

Humic and fulvic acids are also excellent foliar fertilizer carriers and activators. Application of humic acids or fulvic acids in combination with trace elements and other plant nutrients, as foliar sprays, can improve the growth of plant foliage, roots, and fruits (Fenn *et al.*, 1995; Clapp, 2001).

Young plant roots, and growing plants are more responsive to applications of humic substances. Actively growing plant tissues are most responsive to applications of humic substances. Younger tissues have active transport mechanisms that move the required nutrients to sites of metabolic activity. Actively growing plant parts involved in cell divisions and other growth processes readily integrate various trace minerals and growth regulating compounds into on-going metabolic processes. In contrast older plant parts in which metabolic processes have slowed are unable to efficiently utilize added humic substances and associated nutrients.

The uptake of high molecular weight humic acids by roots is primarily passive while the uptake of smaller fulvic acids polymers is primarily metabolic. After humic acids and fulvic acids attain a particular concentration in the root, a fraction (5-30 %) of the total concentration is transported into the shoots and leaves. Radioactive carbon studies indicate that greatest concentration of humic substances accumulates in plant cell walls and cellular organelles, such as the mitochondria and ribosomes. Other similar experiments using radioactive carbon labelled humic and fulvic acids indicate that low molecular weight fulvic acids are much more active compared to high molecular weight humic acids. However, some metabolic reactions may require low concentrations of humic acids in combination with fulvic acids.

Direct effect of humic and fulvic acids on the permeability of plant cell membranes has also been reported which has favourable influence on the transport of various mineral nutrients to sites of metabolic need. Humic substances influence both hydrophilic and hydrophobic sites on the membrane surfaces. In addition, many scientists believe that the phospholipid components of the membranes are electrically altered by humic substances, as a result of which the membrane surface becomes more active in the transport of trace minerals.

Presence of humic substances has been reported to accelerate energy metabolism and enhance chlorophyll content of plant leaves (Nardi *et al.*, 1996; Marschner, 1999). Increase in chlorophyll concentration brings about a correlated increase in the growth and development of plants. Increase in the concentration of m-RNA in response to the presence of humic substances has also been reported. Activation of several such biochemical processes results in an increase in enzyme synthesis and protein content of leaves. Some of the enzymes which are reported to increase are, catalases, peroxidases, diphenoloxidases, polyphenoloxidases, and invertase.

Studies have shown that some molecular components of humic substances act to regulate plant growth hormones. For instance, both humic and fulvic acids inhibit IAA-oxidase, thereby hindering IAA destruction. When IAA is protected from IAA degrading enzymes, IAA continues to stimulate growth processes. Humic substances also increase production of ATP within plant cells. Besides, they provide many "active sites" which function as electron donors because of which humic substances exert positive effects on seed germination, root initiation and plant growth in general. Effect of humic substances on the growth of plants is also because of their influence on the free radical content. Free radicals contain one or more unpaired electrons, are highly reactive, short lived, and capable of participating in many different reactions. The free radical content of humic substances is related to the humification state of the humic substance and greater the humification (low H:C ratios) the darker the colour of the humus. Thus,

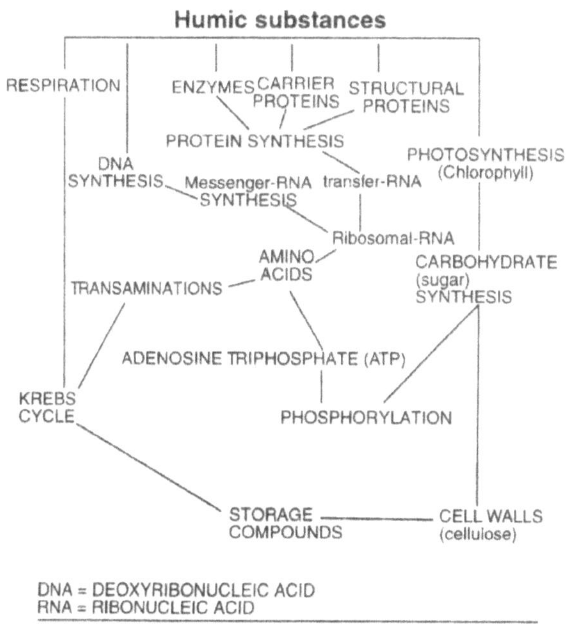

Fig. 7.11. Effect of humic substances on metabolic activities of plants.

humic acids contain a higher free radical content compared to fulvic acids which have a high H:C ratio. The relatively low free radical content of fulvic acids associated with high H:C ratios is indicative of a low degree of chemical condensation for these substances.

Because of the above stated benefits, the worldwide usage of humic substances is extensive (Ghabbour and Davies, 2001). They are an integral part of all ecosystems and play an important role in global cycling of nutrients and carbon (MacCarthy, 2003). A diagrammatic representation of the multifarious direct effects of humic substances on the growth and development of plants is given in Fig. 7.11.

•••

CHAPTER 8

NUTRIENT CYCLING

Nutrient cycling may be defined as the transformation of nutrients from one chemical form to another, and/or the flux of nutrients between organisms, habitats, or ecosystems (Vanni, 2002). Decomposition of detritus is the key process in the nutrient cycling of both terrestrial and aquatic ecosystems (Grimm *et al.*, 2003). Mineralization, the net release of mineral nutrients with the decay of organic matter, is often considered as a three—stage process: first, nutrients in soluble form are leached from the litter; second, immobilization of nutrients; and, finally, net nutrient mineralization from the detritus, thereby making nutrients available for plant uptake (Aerts and Chapin, 2000). However, the relationship between nutrient content in live tissues and in litter is dependent on resorption (Rejmankova, 2005). Nutrient resorption affects litter quality and it consequently affects decomposition rates and nutrient availability. Thus, knowledge of resorption patterns and their determinants is of paramount importance for understanding the role of plants in ecosystem nutrient cycling.

8.1. Nutrient resorption

It refers to the removal of nutrients from senescing plant tissues and their transport to storage organs or growing tissues (Killingbeck, 1986; Aerts, 1996). Terms, such as 'nutrient (re)translocation' or 'nutrient re-absorption' are also used for this process. Resorption may occur throughout a leaf's life (senescing organ), particularly as leaves become progressively shaded (Ackerly and Bazzaz, 1995), but a major pulse of resorption normally occurs shortly before leaf abscission. It is an integral part of the highly ordered process of senescence and appears to occur in most species (Wright and Westoby, 2003). Nutrient resorption from senescing organs is a major nutrient-conservation mechanism (Aerts, 1990) with implications at both the population and the ecosystem level.

The nutrients that are resorbed during senescence are directly available for further plant growth, which makes a species less dependent on current nutrient uptake (Aerts and Chapin, 2000). Thus, at ecosystem level nutrient resorption from senescing leaves has profound influence on mineral cycling. Nutrients that are not resorbed, however, are recycled through litterfall when is decomposed and the nutrients contained in the litter are mineralised and made available for uptake by plants and microbes. Most of the information regarding resorption concerns nitrogen (N) and phosphorus (P) because most terrestrial ecosystems in temperate regions are thought to be limited by N, while many tropical terrestrial ecosystems are P limited (Vitousek *et al.*, 1997).

Much of the importance that nutrient resorption has gained over the years is largely due to the studies conducted by Aerts (1996) and Killingbeck (1996). While the former summarized published data on N- and P-resorption efficiency (NRE, PRE) from the senescing leaves of perennial plants grouped into evergreen shrubs and trees, deciduous shrubs and trees, graminoids, and forbs, the latter analysed patterns of N and P content in deciduous and evergreen woody perennials. Killingbeck (1996) introduced the concept of resorption proficiency (terminal content of a nutrient in senescent leaves) and differentiated between complete and incomplete resorption.

A commonly used parameter to quantify resorption is resorption efficiency. This parameter quantifies the percentage of the nutrient pool withdrawn from the foliage before abscission (functional disconnection of leaves that remain attached to the plant), and is determined by measuring the nutrient pools of mature and abscised leaves. The nutrient pool is usually expressed on the basis of leaf mass or leaf area.

Growth form and resorption

Resorption efficiency in respect of nitrogen and phosphorus recorded in plants of different growth forms is presented in Table 8.1. As is evident from these data, nitrogen concentration in mature leaves of evergreen shrubs and trees and of graminoids is lower than in deciduous shrubs, trees and forbs. A similar pattern is observed for P concentration in mature leaves. Mean N and P resorption from senescing leaves is 50% and 52%, respectively. Thus, on an average about half of the N and P contained in leaves can be re-used after senescence of the leaves. There are, however, no differences in N and P resorption between various growth-forms (Aerts and Chapin, 2000) which implies that nutrient resorption, though an important nutrient conservation strategy at the species level, apparently does not explain the distribution of growth-forms in habitats differing in soil fertility. For example, species from nutrient-poor sites (evergreens) have not adapted to low soil fertility by having high nutrient resorption efficiency (Aerts, 1996; Killingbeck, 1996).

Table 8.1. Mean (± SD) nitrogen (N) and phosphorus (P) concentration and resorption efficiency in mature leaves of perennial plants.

Growth forms	**Concentration** (mg g^{-1})		**Resorption efficiency** (%)	
	N	P	N	P
Evergreen shrubs and tress	13.7±5.2 (95)	1.02±0.56 (74)	46.7±16.4 (108)	51.4±21.7 (88)
Deciduous shrubs and trees	22.2±7.4 (95)	1.60±0.92 (78)	54.0±15.9 (115)	50.4±19.7 (98)
Forbs	22.2±5.8 (29)	1.86±0.72 (14)	41.4±21.4 (33)	42.4±30.3 (18)
Graminoids	16.0±4.0 (22)	0.95±0.51 (13)	58.5±14.2 (31)	71.5±16.0 (22)
All growth forms(taken together)	18.3±7.4 (241)	1.33±0.81 (179)	50.3±17.3 (287)	52.2±22.1 (226)

- Number of observations are given in parentheses.
- Modified from Aerts, (1996)

Nutritional controls of resorption

Aerts (1996) carried out analysis of published data obtained from various fertilization experiments covering 60 species (Table 8.2) in order to investigate intraspecific differences in N and P resorption in relation to site fertility. No response in N resorption was noticed in 63% of the experiments analysed, whereas in 32% experiments decrease in N resorption was observed upon increase in nutrient availability. Of the 37 species analysed for P resorption, 57% species showed no response and in 35% of species P resorption decreased in response to enhanced nutrient supply. Evergreen shrubs and trees especially showed low response in nutrient resorption to altered nutrient availability.

Resorption proficiency

Another parameter used to quantify leaf nutrient resorption is resorption proficiency which was introduced and defined by Killingbeck (1996) as the level to which a plant has reduced an element in its senescing leaves. Species with high nutrient resorption proficiency thus show low nutrient levels in litter. The minimum level to which a plant can potentially reduce its nutrients is about 3 mg g^{-1} dry weight for N and between 0.7 and 1 mg g^{-1} dry weight for P (Aerts and Chapin, 2000; Côté *et al.*, 2002). Since this level is not always attained, Killingbeck (1996) also defined litter N and P levels that characterize complete, intermediate and incomplete resorption for deciduous and evergreen woody perennials. Some external factors that can directly lead to incomplete leaf nutrient resorption are for instance frost, which prematurely arrests the resorption process (Norby *et al.*, 2000), and strong wind, which can prematurely detach leaves from the plant (Oland, 1963; Killingbeck, 1988).

These factors may contribute to the variation in nutrient resorption proficiency among different years and regions. Other external factors that correlate with the level of N and P resorption proficiency are latitude (Berg *et al.*, 1995), N and P availability (Shaver and Melillo, 1984; Kemp *et al.*, 1994; Eckstein *et al.*, 1999) and temperature (Nordell and Karlsson, 1995). In addition, intrinsic growth form differences may also be responsible for variation in N and P resorption proficiency. For instance, plant species harbouring nitrogen-fixing symbionts are a clearly distinct group, which show lower N resorption proficiency compared to other species.

Lack of nutritional controls on nutrient resorption has led to advancement of several other possible controls, including the relative sink strength of plant organs (Nambiar and Fife, 1991), the rate of phloem transport (source-sink interactions) (Chapin and Moilanen, 1991) and soil moisture availability (Boerner, 1985; Pugnaire and Chapin, 1993).

Table 8.2. Enhanced nutrient supply and phenotypic response to nutrient resorption.

Nutrient	Growthform	Response of percentage number of species to increased nutrient availability			Number of species examined
		Increase	Decrease	None	
Nitrogen	Evergreen shrubs and trees	4	21	75	24
	Deciduous shrubs and trees	0	42	58	12
	Forbs	20	0	80	10
	Graminoids	0	64	36	14
	All growth forms (taken together)	5	32	63	60
Phosp horus	Evergreen shrubs and trees	11	16	19	19
	Deciduous shrubs and trees	0	62	19	8
	Forbs	17	33	8	6
	Graminoids	0	75	4	4
	All growth forms (taken together)	8	25	57	37

Modified from Aerts, (1996)

Biochemistry of resorption

Plant leaves, upon their entry into the senescence phase at the end of their life span, undergo many changes that assist the retention of valuable nutrients like N and P. Efficient resorption of N from senescing leaves is accompanied by marked de -greening (Thomas *et al.*, 2002). It is due to the breakdown of photo-destructive chlorophyll to colourless breakdown products, which unmasks the partially retained carotenoids and gives senescent leaves their typical yellow colour (Thomas, 1997; Matile *et al.*, 1999). The highly regulated chlorophyll breakdown is necessary to gain access to the large pool of N-containing proteins associated with chlorophyll. These proteins are subsequently hydrolysed and modified to N-rich amino acids prior to their loading onto the phloem (Feller and Fischer, 1994). In the meanwhile, P containing DNA, RNA and phospholipids are also broken down (Matile, 1997). The energy required for these activities is supplied by the relatively unstable, but still active photosynthetic apparatus (Hoch *et al.*, 2001). As excess light is particularly harmful at this stage, some species turn their leaves deeply red with anthocyanins as a photo-protective mechanism (Feild *et al.*, 2001; Hoch *et al.*, 2001). But with the collapse of intercellular organisation and turgor, phenol oxidases from the plastids can contact the phenolics that had been stored in the vacuole, leading to the formation of the brown melanin (Matile, 2000). At this stage, hydrolases, which had been accumulating during the late stage of leaf senescence, are also released from the vacuole. Their function remains illusive and is not likely to be related to nutrient conservation (Matile, 1997). To ensure the ordered dehiscence of the old leaf, many species also form an abscission layer during senescence. Despite the beneficial aspects of nutrient resorption for the plant, the nutrient pool from leaves is never resorbed totally (Killingbeck, 1996). The resorption machinery itself consists of proteins, and is logically not able to dismantle itself without direct loss of function. The same line of thinking applies to the phospholipids from membranes, as there is probably a minimal amount of phospholipids necessary for a cell to remain intact. In addition, the colourless N containing chlorophyll catabolites, which are dumped in the vacuole, also contribute to the N in litter. As a result, average resorption efficiency of N and P is only about 50% ± 20 (SD) (based mostly on shrubs and trees) during leaf senescence (Aerts, 1996).

8.2. Nutrient leaching

Very few quantitative estimates of the leaching of nutrients from fresh litter are available (Aerts and Chapin, 2000) though its significance has been emphasized by many workers. Swift *et al.* (1979) reported that nutrient ions like K and Mg are very mobile and easily leached from fresh litter, whereas N, P and Ca show a much lower rate of leaching. Most of the studies related to quantification of leaching of nutrients of senesced leaves concern N and P. Leaching, estimated as a percentage of the initial N and P content of the senesced leaves, reveals a highly variable pattern (Table 8.3).

Influence of site fertility on leaching in four deciduous species was investigated by Boerner (1984) who found lower N and P leaching at low-fertility sites in 10 out of 16 cases. Though no consistent effect of fertilization on initial N content of leaf litter of *Schizachyrium scoparium* was apparent but significant positive relationship was noticed between N leaching and initial N content of the leaves (Pastor *et al.*, 1987). Similar studies carried out by Aerts and De Caluwe (1997) on *Carex* species revealed relatively higher N and P leaching losses from litters with high N and P concentration than low-nutrient litter.

8.3. Nutrient mineralization

The feedback between detritus and nutrient cycling processes plays a major role in the regulation of nutrient availability and net primary productivity in ecosystems (Chapin, 1980; Vitousek, 1982; 1984). While several studies have examined site-specific feedbacks between litter chemistry and nutrient availability, little is known about the interaction between climate, litter chemistry, soil organisms and nutrient availability across different ecosystems (Scott and Binkley, 1997). Nutrient release, which is influenced by litter quality, affects ecosystem functioning by changing the balance between nutrient release and uptake by decomposers:Litter quality, in turn, is'determined by the nature of plant species and/or functional groups comprising a particular ecosystem (Semmartin *et al.*, 2004). The availability of below-ground resources is also a major ecological control over litter quality. For example, nutrient rich habitats support plants with high growth rates, high leaf nutrient content, low nutrient use efficiency and low levels of herbivore defence while opposite holds true for plants that grow in nutrient poor habitats (Chapin, 1980; Aerts and Chapin, 2000; Coley *et al.*, 1985). Such rapidly growing plants from high resource sites typically produce litter that decomposes quickly. Besides, plants in nutrient poor habitats are likely to produce less litter with lower nutrient concentration than plants in nutrient rich habitats. Also the concentration of different nutrients in litterfall varies according to the availability of the nutrients to the plants (Vitousek and Sanford, 1986).

Table 8.3. Variation in the extent of N and P leaching from senescing leaves.

Species	Location	N leaching (%)	P leaching (%)	Reference
Molin caerulea	England	0	0	Morton, 1977
Four deciduous and evergreen trees	Alaska	0.1-0.3	0.04-0.6	Chapin and Kedrowski, 1983
Four deciduous tree species	Ohio	0-8	0-10	Bonner, 1984
Schoenus ferrugineus	Germany	0	0	Ganzert and Pfadenhauer, 1986
Schizachyrium scoparium	Minnesota	4-57*	ND	Pastor et *al.*, 1987
Betula papyrifera	Alaska	31	31	Chapin and Moilanen, 1991
Seven evergreen and deciduous species	Mediter-ranean	0-51	ND	Ibrahima et *al.*, 1995
Four Carex species in fens	The Neth erlands	8-12	31	Aerts and DeCaluwe, 1997

* Four levels of N supply; N leaching positively related to initial N content.
ND., not determined

Source: Aerts and Chapin (2000)

Notwithstanding the variations in its elemental composition, litter constitutes an important nutrient reservoir in an ecosystem and nutrients contained in the litter are released upon its decomposition. Thus, it is imperative to have an insight into broad biogeographic patterns of nutrients, particularly nitrogen (N) and phosphorus (P) in the litter, because the same typically reflects their availability in soil. In this context, the global study carried out by Reich and Oleksyn (2004), based on a data set including 5,087 observations of leaf nitrogen (N) and phosphorus (P) for 1,280 plant species at 452 sites varying in latitude is noteworthy. This global data set was analysed in three different ways: (i) using all data (i.e., treating all observations equally), (ii) averaging by species, and (iii) averaging by species within temperature ranges (called as bins). In general, the study reveals that leaf N and P contents decline and the N/P ratio increases towards equator with increase in average temperature and growing season. Such a pattern was also shared by the five dominant plant groups, namely, coniferous trees and angiospermous grasses, herbs, shrubs, and trees (Table 8.4). This study further reveals:

a. increase in leaf N and P content from tropics to the cooler and drier mid-latitudes because of temperature-related plant physiological stoichiometry and biogeographical gradients in soil substrate age. At high latitudes, leaf N and P may plateau or decrease because of cold temperature effects on biogeochemistry, and,

b. increase in the N/P ratio with increase in mean temperature towards the equator, because P is a major limiting nutrient in older tropical soils and N is the major limiting nutrient in younger temperate and high-latitude soils.

Despite the fact that annual carbon and nutrient inputs to the soil from fine roots frequently equal or exceed those from leaves, they are poorly represented in global models. Taking note of it Jackson *et al.* (1997) provided a global budget for fine root biomass, length and surface area vis-à-vis soil depth and global estimates of nutrient pools in fine roots. Total fine root biomass ranges from 0.27 $kg.m^{-2}$ in deserts to 1.5 $kg.m^{-2}$ in temperate grasslands (Table 8.5).

Table 8.4. Relationship of leaf N, P, and N/P ratio with mean annual temperature (MAT) and absolute value of latitude (AL).

Attributes	**Equation**	r^2	***P value***
N vs. MAT, bins	log y = 1.25891 + 0.00863x - 0.00046x^2	0.75	0.008
N vs. MAT, all data	log y = 1.27373 + 0.00595x - 0.00036x^2	0.05	<0.0001
P vs. MAT, bins	log y = 0.35097 + 0.01518x - 0.00366x^2 + 0.00009x^3	0.95	0.0002
P vs. MAT, all data	log y = 0.37413 - 0.00901x - 0.0005x^2	0.39	<0.0001
N/P vs. MAT, bins	log y = 0.900 + 0.014x	0.85	0.0001
N/P vs. MAT, all data	log y = 0.88442 + 0.01547x	0.31	<0.0001
N vs. AL, bins	log y = 1.06944 + 0.00978x - 0.00011x^2	0.69	0.02
N vs. AL, all data	log y = 1.17942 + 0.00231x	0.04	<0.0001
P vs. AL, bins	log y = -0.4506 + 0.07227x - 0.0051x^2 + 0.00014x ^{3} - 0.000118x4	0.98	0.001
P vs. AL, all data	log y = -0.3246 + 0.0122x	0.34	<0.0001
N/P vs. AL, bins	log y = 1.30985 - 0.00377x - 0.00006x^2	0.78	0.005
N/P vs. AL, all data	log y = 1.37432 - 0.00769x	0.24	<0.0001
Data by groups			
N vs. MAT			
Grasses	log y = 1.3374 - 0.000004x - 0.00056x^2	0.10	0.0035
Herbs	log y = 1.22296 + 0.02809x - 0.00126x^2	0.08	<0.0001
Shrubs	log y = 1.24046 + 0.00445x - 0.00041x^2	0.07	0.0005
Trees	log y = 1.33628 + 0.00007x - 0.00017x^2	0.05	<0.0001
Conifers	log y = 1.12038 - 0.00431x	0.10	0.009

Contd...

P vs. MAT			
Grasses	$\log y = 0.34277 - 0.00296x - 0.00114x^2$	0.31	<0.0001
Herbs	$\log y = 0.43293 + 0.00442x - 0.00108x^2$	0.35	<0.0001
Shrubs	$\log y = 0.29516 + 0.00107x - 0.00128x^2$	0.48	<0.0001
Trees	$\log y = 0.3604 - 0.02022x$	0.25	<0.0001
Conifers	$\log y = 0.23425 - 0.01433x$	0.49	<0.0001
N/P vs. MAT			
Grasses	$\log y = 0.95566 + 0.01014x$	0.11	0.001
Herbs	$\log y = 0.79681 + 0.0172x$	0.24	<0.0001
Shrubs	$\log y = 0.91688 + 0.01623x$	0.25	<0.0001
Trees	$\log y = 1.00264 + 0.01182x$	0.20	<0.0001
Conifers	$\log y = 0.88186 + 0.01074x$	0.36	<0.0001
Data by groups			
N vs. P			
Conifers	$\log y = 1.0398244+0.3666713x$	0.31	<0.0001
Grasses	$\log y = 1.1462143+0.5112318x$	0.46	<0.0001
Herbs	$\log y = 1.1443574+0.4652923x$	0.38	<0.0001
Shrubs	$\log y = 1.1519265+0.2975617x$	0.40	<0.0001
Trees	$\log y = 1.2213062+0.4507624x$	0.58	<0.0001
Data by group (5, bins)			
N vs. MAT			
Grasses	$\log y = 1.37501 - 0.00036x+7^2$	0.69	0.006
Herbs	$\log y = 1.395 - 0.0002269\ x + 7^2$	0.47	0.03
Shrubs	$\log y = 1.221 +0.006155x - 0.0004992x^2$	0.57	0.05
Trees	$\log y = 1.315 + 0.004906x - 0.0003456x^2$	0.96	<0.0001
Conifers	$\log y = 1.332 - 0.0001156\ x + 7^2$	0.72	0.002

Contd...

P vs. MAT			
Grasses	$\log y = 0.40211 - 0.00075\,x + 7^2$	0.81	0.001
Herbs	$\log y = 0.530 - 0.0005443\,x + 7^2$	0.76	0.002
Shrubs	$\log y = 0.282 + 0.005518x - 0.001418x^2$	0.89	0.0004
Trees	$\log y = 0.332 - 0.000447x + 7^2$	0.76	0.0009
Conifers	$\log y = 0.232 - 0.0003544\,x + 7^2$	0.86	<0.0001
N/P vs. MAT			
Grasses	$\log y = 0.9133 + 0.00039x+7^2$	0.68	0.006
Herbs	$\log y = 0.803 + 0.014x$	0.83	0.0006
Shrubs	$\log y = 0.902 + 0.016x$	0.67	0.004
Trees	$\log y = 0.979 + 0.012x$	0.69	0.003
Conifers	$\log y = 0.89272 + 0.0103x$	0.78	0.0007

Modified from Reich and Oleksyn (2004)

Table 8.5. Average total and live fine root biomass in different biomes.

Biomes	**Total fine root biomass** (kg.m^{-2})	**Live fine root biomass** (kg.m^{-2})
Boreal forest	0.60	0.23
Desert	0.27	0.13
Sclerophyllous shrub and trees	0.52	0.28
Temperate coniferous forest	0.82	0.50
Temperate deciduous forest	0.78	0.44
Temperate grassland	1.51	0.95
Tropical deciduous forest	0.57	0.28
Tropical evergreen forest	0.57	0.33
Tropical grassland/savannah	0.99	0.51
Tundra	0.96	0.34

Abridged from Jackson *et al.* (1997); Fine roots are defined as those with diameter of ≤ 2 mm.

Average live fine root biomass in rest of biomes ranges from 0.1 to 0.5 kg.m^{-2}. Based on this data, a global estimate of total root biomass, fine root biomass and live fine root biomass (Table 8.6) and a global budget for nutrient concentrations in fine roots (Table 8.7) is computed. Tropical grasslands/savannahs, temperate grasslands and tropical rainforests have the greatest total fine root biomass ($\approx 1\times10^{16}$g for each) and globally there are 7.8×10^{16}g of fine roots. The carbon in this global fine root pool is approximately 5% of the size of the atmospheric carbon pool. The global pool of N in living fine roots was 4.8×10^{8} Mg, approximately one-seventh of the estimate for all terrestrial vegetation and the average C:N ratio of live fine roots globally is 42.

Table 8.6. Global land area, total root biomass, total fine root biomass and live fine root biomass in different biomes.

Biome	**Land area** (10^6 km^2)	**Total root biomass** (10^9 Mg)	**Total fine root biomass** (10^9 Mg)	**Live fine root biomass** (10^9 Mg)
Tropical rainforest	17.0	83	9.7	5.7
Tropical seasonal forest	7.5	31	4.3	2.1
Temperate evergreen forest	5.0	22	4.1	2.5
Temperate deciduous forest	7.0	29	5.6	3.1
Boreal forest	12.0	35	7.2	2.8
Woodland and shrubland	8.5	41	4.4	2.4
Savannah	15.0	21	14.9	7.7
Temperate grassland	9.0	14	13.6	8.5
Tundra/alpine	8.0	10	7.7	2.7
Desert	18.0	6.6	4.9	2.3
Cultivated	14.0	2.1	2.1	1.1
Total	**121**	**292**	**78.2**	**40.8**

Source: Abridged from Jackson *et al.* (1997)

Table 8.7. Global nutrient pools in living fine roots, and total fine roots.

Element	Concentration in living fine roots (%)	Nutrients in living fine roots (10^9 Mg)	Nutrients in total fine roots (10^9 Mg)
C	48.8	19.9	38.1
N	1.17	0.48	0.92
P	0.11	0.044	0.085
K	0.30	0.12	0.24
Ca	0.41	0.17	0.32
Mg	0.14	0.054	0.11
S	0.088	0.036	0.069

Abridged from Jackson *et al.* (1997)

Decomposition of dead organic matter completes the intrasystem cycle by releasing nutrient elements for plant uptake through the process of mineralization. These decomposition/ mineralization processes are mainly mediated by bacteria and fungi via extracellular enzymes, although larger organisms (invertebrates) also contribute significantly to nutrient cycling (Vanni, 2002). However, different nutrients exhibit different release and cycling rates from the detritus. For example, nutrients, such as K which is easily leached from litter, may show more rapid turnover rates than total organic matter. On the contrary, nutrients, such as N which are subject to immobilization, turnover more slowly than bulk organic matter. It is generally accepted that lower carbon to nitrogen ratios promote nutrient release whereas higher ratios induce microbial immobilization. Thus, different nutrients show variable residence time in the litter (Table 8.8).

Organic matter mineralization, and thereby availability of nutrients, is influenced by species characteristics, such as litter quality. Higher litter lignin concentrations generally retard litter decay rates, and are related to decreased N availability. However, factors such as climate and stand age (in forests) may also influence N availability. A comparative study of annual net N mineralization across North American forest sites carried out by Scott and Binkley (1997) revealed that litterfall quantity, N concentration and N content correlate poorly with net N mineralization across the range of sites (Table 8.9). However, litter lignin:N ratio explained more of the variation in net N mineralization than climatic factors.

Table 8.8. Mean residence time for organic matter (OM) and nutrients in the surface litter of forest and woodland ecosystems*.

Ecosystem type	**Mean residence time** (yr)					
	OM	N	P	K	Ca	Mg
Boreal forest	353	230	324	94	149	455
Temperate coniferous forest	17	17.9	15.3	2.2	5.9	12.9
Temperate deciduous forest	4	5.5	5.8	1.3	3.0	3.4
Mediterranean forest	3.8	4.2	3.6	1.4	5.0	2.8
Tropical rainforest	0.4	2.0	1.6	0.7	1.5	1.1

*Values are computed by dividing the forest floor mass by the mean annual litterfall.

Table 8.9. Lignin and N content of litter and annual net N mineralization.

Site	***Species***	**Lignin** (g kg $^{-1}$)	**N** (g kg $^{-1}$)	**Net N mineralization** (kg ha^{-1} year^{-1})
Rocky Mountain Region	*Populus tremuloides*	176	8.2	24.3
		182	5.9	26.3
		195	6.1	21.1
		200	6.6	7.6
		203	6.0	11.9
		193	6.9	29.9
		198	10.2	25.9
		188	7.8	51.4
		196	7.1	27.4
		210	7.8	13.4
	Pinus contorta	244	4.8	2.1
		250	4.8	5.6

Contd...

		256	4.4	1.5
		237	5.5	4.1
		263	4.3	3.1
		251	4.0	1.5
		243	4.8	-4.6
		250	4.4	-0.6
		261	4.6	2.1
		273	3.8	1.4
	Abies lasiocarpa	295	6.3	8.9
		283	4.7	4.1
		262	6.7	5.8
		249	6.6	12.2
		276	6.7	9.0
		238	5.0	4.6
		269	5.6	0.7
		274	5.4	11.2
		266	5.1	2.1
		244	6.2	4.4
Cedar Cre	*Agrostis scabra*	51	10.1	123.6
Natural History Area	*Agropyron repens*	61	8.7	40.0
	Poa pratensis	39.8	7.1	28.0
	Schizachyrium scoparium	57.5	4	9.8
	Andropogon gerardi	50.8	3.3	17.8
Coullee Experimental Forest	*Larix decidua*	126	10.7	117.0
	Pinus strobus	225	8.2	87.0
	Pinus resinosa	219	7.2	51.0
	Picea abies	268	9.9	46.0
	Quercus rubra	248	8.6	55.0

University of Wisconsin Arboretum	*Quercus velutina*	129	8.1	125.0
	Quercus rubra	143	8.5	135.0
	Quercus alba	67	7.8	94.0
	Acer saccharum	133	8.3	99.0
	Pinus strobus	204	7.3	81.0
	Pinus resinosa/ strobus	212	5.8	61.0
	Picea abies	230	10.4	58.0
Harvard Forest	*Pinus resinosa*	220	7.1	80.0
	Mixed hardwoods	180	9.1	70.0
Eli Whitney Forest	*Fraxinus penn-sylvanica*	183	10.7	41.1
	Pinus strobus	173	11.5	83.7
	Picea abies	159	11.9	58.5
Blackhawk Island	*Pinus resinosa*	228	6.9	28.0
	Pinus strobus	202	8.3	52.0
	Quercus rubra	166	9.6	78.0
	Quercus rubra	151	10.1	86.0
	Acer saccharum	144	10.1	98.0
Blodgett Research Forest	Old-growth mixed conifer forest	261	5.7	29.0
Isle Royale National Park	Deciduous (protected)	293	9.7	38.3
	Coniferous (browsed)	273	9.8	23.8

Source : Modified from Scott and Binkley (1997)

Other studies (Moro and Domingo, 2000) have also reported that nutrient content in litter (mainly N) regulates the early stages of decomposition (1-2 years) whereas later stages appear to be regulated by the percentage of lignin and other resistant components (Berg and Ekbohm, 1983). A 'critical C:N ratio' for mineralization seems to be between 20:1 to 30:1 (Lutz and Chandler, 1946). However, other authors have pointed out that this critical index changes greatly depending on a wide variety of factors (Stohlgren, 1988). In conifers critical C:N ratios ranging froth 109 to 19 have been reported (Berg, 1986). Since N is an essential element strongly retained by plants, so freshly fallen litter normally has a higher C:N ratio than the critical threshold for mineralization, especially in nutrient-poor coniferous or sclerophyllous Mediterranean species (Moro and Domingo, 2000). Different levels of N during the initial stages impact the mineralization and immobilization of nitrogen in the decomposing litter. However, studies have revealed that most of the nitrogen is always incorporated into the soil organic matter, and its decomposition primarily determines nitrogen mineralization and thereby plant-available nitrogen and primary productivity (Knops *et al.*, 2002). Rates of gross nitrogen mineralization, which are often an order of magnitude higher than net mineralization, indicate that nitrogen cycling within ecosystems is dominated by a microbial nitrogen loop. Nitrogen is released from the soil organic matter and incorporated into microbial biomass. Upon their death, the nitrogen is again incorporated into the soil organic matter, However, this microbial nitrogen loop is driven by plant-supplied carbon and provides a strong negative feedback through nitrogen cycling on plant productivity. Evidence supporting this hypothesis is strong for temperate grassland ecosystems whereas data for other terrestrial ecosystems is too limited to draw any definite conclusions. Furthermore, a significant inverse correlation between N concentration and percentage of mass remaining seems a general trend, since it has been described by many workers in a large variety of species and can occur with or without release of N (Carlyle, 1986). Eventually, enough C is lost and the C:N ratio of litter becomes equivalent to that of bacterial and fungal tissues. The relatively rapid respiration of C and the immobilization of N by microbial tissue produces a decrease in the C:N ratio over time.

Though much attention has been devoted to determine biological and environmental factors influencing the release and mineralization of N from plant litter (Salas *et al.*, 2003), relatively few studies have evaluated the biological transformations associated with the release of P from residues and subsequent accumulation and turnover of organic P during decomposition (Dalai, 1979; McLaughlin *et al.*, 1988b; Umrit and Friesen, 1994). Notwithstanding the differences among species in the behaviour of phosphorus during the decomposition process (Gosz *et al.*, 1973; Lousier and Parkinson, 1978; Schlesinger, 1985; Stohlgren, 1988), use of radioactive tracers (^{32}P and ^{33}P) reveals that P release from residues is relatively rapid (Friesen and Blair, 1988). Phosphorus immobilization by microorganisms, turnover of microbial P, and mineralization of microbial by-products seem to be the major processes regulating cycling and availability of P (McLaughlin *et al.*, 1988a). Recent decomposition studies using litter bags indicate that a significant part of microbial immobilization of P takes place inside decomposing plant residues (Schomberg and Steiner, 1999 and Mafongoya *et al.*, 2000) where microbial turnover and P mineralization might differ from the processes taking place in the bulk soil. Phosphorus immobilization in decomposing residues is believed to be responsible for as much as 70% increase in the amount of P in agricultural residues (Tian *et al.*, 1992; Schomberg and Steiner, 1999), and from 20% to 200% increase in tree leaves (Conn and Dighton, 2000). Some reports suggest that immobilization of P in residues could significantly decrease P availability from added plant residues (Mafongoya *et al.*, 2000). Possible role of soil fungi suggested in the immobilization (Salas *et al.*, 2003) could be particularly important in acid soils where there is often a predominance of fungi compared with bacterial populations. In addition to mediating the turnover of P, soil microorganisms may also constitute a significant reservoir of P (Brookes *et al.*, 1984; Tiessen *et al.*, 1994). Soil microbial biomass shows strong seasonal variations which are obviously related to the quantity of labile carbon, moisture and temperature of soil (Dalal, 1998).

While N and P are generally the two elements that limit primary productivity in most terrestrial and aquatic ecosystems, but potassium (K), calcium (Ca), and magnesium (Mg) are also essential macronutrients for energy metabolism, photosynthesis, and membrane transport of plants (Slovic, 1997) and litter is a major source of these nutrients as well (Likens and Bormann, 1995). Study of Hasegawa and Takeda (1997) on the dynamics of K, Ca, and Mg during litter decomposition has revealed that K leaches out quickly from decomposing litters, mineralization of Ca decreases with loss of carbon during litter decomposition, and Mg often shows an intermediate release pattern. Osono and Takeda (2004) also described dynamics of K, Ca, and Mg in cool temperate forests of Japan and reported that initial concentration (C_0) of K and Mg in the litter and the critical concentrations ($C_{minimom}$ or $C_{maximom}$) of Ca and Mg are best predictors of the pattern of change in concentrations of these elements during decomposition. They found that the final concentrations are not dependent on the initial concentration of the element in the litter but instead on site conditions. Since K is not a structural material and exists mainly in solution in plant cells, it leaches out quickly from decomposing litters. The initial phase of K leaching is thus correlated to the initial concentration while late phase of K dynamics is characterized by seasonal changes (Osono and Takeda, 2004). Dynamics of Ca in decomposing litter shows either increase and decrease or the decrease phase only (Osono and Takeda, 2004). Previous studies have also observed the two-phase pattern (Attiwill 1968; Hasegawa and Takeda 1997; Bhatta *et al.*, 2000) or the onephase pattern (MacLean and Wein, 1978; Edmonds, 1984; Adams and-Angradi 1996). Concentration of Ca initially in the litter does not predic the pattern of change in its concentration during the entire process of decomposition. This is attributed to covalent bonding of Ca to pectin within the middle lamella in litter. Thus, leaching or decomposition of more readily available organic components may cause relative increase in Ca concentration of the decomposing litter independent of initial Ca content. On the other hand, in the late phase, the release of Ca is dependent on decomposition of the structural components. Dynamics of Mg in decomposing litter shows two phases; the initial leaching phase and the late immobilization phase (Osono and Takeda, 2004). This pattern of change in Mg concentration has been reported in previous studies also (Attiwill, 1968; Edmonds, 1984; Berg *et al.*, 1987; Blair, 1988; Laskowski and Berg, 1993; Hasegawa and Takeda, 1997; Salamanca *et al.*, 1998).

In conclusion it needs to be emphasized that aquatic and terrestrial biogeochemists share the view that the C:N ratio of decomposing material influences rates of decomposition and consequently nutrient release in both forests as well as streams. Though there has been little explicit comparison of decomposition rates in aquatic and terrestrial systems (Grimm *et al.*, 2003), yet in a stream or on the forest floor decomposition of litter with high C content and low N content is slower

Table 8.10. Comparison of nutrient returns in litter from forests around the world.

Country and location	**Forest type**	**Leaf litterfall** (t $he^{-1}y^{-1}$)	**N** (mg g^{-1})	**P** (mg g^{-1})	**K** (mg g^{-1})	**Ca** (mg g^{-1})	**Mg** (mg g^{-1})	**Authors**
Cameroon, Korup NP	Ectomycorrhizal semi-evergreen	5.6	17.1	0.640	4.93	12.9	2.39	Chuyong *et al.*, 2000
Cameroon, Korup NP	Non-ectomycorrhizal semi-evergreen	5.0	17.1	0.680	4.34	11.9	2.49	Chuyong *et al.*, 2000
Ivory Coast, Banco (plateau)	Evergreen	8.7	15.0	0.690	2.2	5.6	4.6	Bernhard, 1970[1]
Ivory Coast, Banco (valley)	Evergreen	7.9	18.0	1.600	9.1	9.5	4.1	Bernhard, 1970[1]
Ivory Coast, Yapo (plateau	Evergreen	6.4	14.0	0.500	2.8	13.2	2.9	Bernhard, 1970[1]
Ivory Coast, Yapo (valley)	Evergreen	5.9	14.0	0.530	4.9	13.6	3.2	Bernhard, 1970[1]
Brazil, Manaus	Igapo	5.3	15.0	0.530	2.9	5.3	1.3	Adis *et al*, 1979[1]
Brazil, Manaus	Terra firme	6.4	15.0	0.240.	2.1	4.8	1.8	Franken *et al.*, 1979[1]
Brazil, Manaus	Riverine	4.3	13.0	0.280	2.6	3.5	0.17	Franken *et al.*, 1979[1]

Contd...

Brazil, Manaus	Terra firme	5.6	15.0	0.300	1.8	2.2	1.8	Klinge & Rodriguez, 1968*a*, b[1]
Brazil, Belém	Terra firme	8.0	17.0	0.410	1.7	3.1	2.8	Klinge, 1977[1]
Brazil, Belém	Varzea	7.5	12.0	0.360	2.6	8.7	3.0	Klinge, 1977[1]
Brazil, Belém	!gapo	6.7	12.0	0.360	3.2	8.0	3.8	Klinge, 1977[1]
Brazil, Maraca Island	Evergreen	6.3	13.0	0.580	4.7	7.4	2.7	Scott *et al.*1993
Columbia, Magdalena valley	Evergreen	6.5	13.0	0.350	1.3	8.0	1.9	Fölster & de las Salas, 1976[1]
Columbia, Magdalena valley	Evergreen	6.6	12.0	0.410	3.3	14.0	1.3	Fölster & de las Salas, 1976[1]
Venezuela, San Carlos	Oxisol	7.9	16.0	0.320	2.4	1.7	0.7	Medina & Cuevas, 1989[2]
Venezuela, San Carlos	Caatinga	5.6	7.0	0.500	2.1	7.7	3.1	Medina & Cuevas, 1989[2]
Venezuela, Caparo	Mixed evergreen	6.1	16.0	1.500	7.4	21.0	4.0	Franco, 1979[1]
Malaya, Tasek Bera	Fresh water swamp	7.2	9.1	0.190	3.0	14.0	3.4	Furtado, 1980[1]
Malaya, Pasoh	Dipterocarp	3.4	12.0	0.300	3.8	7.0	2.2	Lim, 1978[1]

Contd...

Sarawak, Gunung Mulu NP	Alluvial	6.6	9.0	0.267	2.6	24.0	2.0	Proctor *et al.*, 1983b
Sarawak, Gunung Mulu NP	Dipterocarp	5.4	9.5	0.105	4.5	1.5	1.1	Proctor *et al*, 1983b
Sarawak, Gunung Mulu NP	Heath	5.6	5.7	0.142	2.3	8.9	1.6	Proctor *et al.*, 1983b
Sarawak, Gunung Mulu NP	Limestone	7.3	11.7	0.376	1.6	31.0	3.3	Proctor *et al.*, 1983b
Sarawak, Gunung Mulu NP	Dipterocarp	6.7	10.0	0.170	4.3	5.1	1.2	Proctor *at al*, 1983 [1]
Sabah, Sepilok	Alluvial diptero-carp	5.9	13.5	0.288	3.80	9.78	1.98	Bagchi, 2002
Sabah, Sepilok	Sandstone ridge dipterocarp	3.8	9.0	0.098	2.05	3.68	2.16	Bagchi, 2002
Sabah, Sepilok	Sandstone valley dipterocarp	4.2	9.0	0.148	2.63	4.46	2.26	Bagchi, 2002
Sabah, Sepilok	Heath	4.0	7.2	0.069	2.20	6.73	2.27	Bagchi, 2002
Australia, Whian Whian	Sub-tropical rainforest	7.4	18.0	1.400	5.6	18	-	Webb *et al.*, 1969[1]

Data from Proctor (1984) [1] , Scott *et al.* (1993)[2] and original papers.

than that of litter with a lower initial C:N ratio. But over the course of decomposition the ratio converges to a lower, more uniform value (Aber and Melillo, 1991; lrons *et al.*, 1994) albeit initial differences in the C:N ratio of litters. This similarity persists even though other factors regulating litter decomposition differ dramatically.

In general, elemental content of detritus coupled with redox conditions dictate rates of detrital decomposition. The elemental stoichiometry of the detrital pool in any ecosystem is a reflection of nutrient ratios of all of the major biotic compartments as influenced by the turnover rates of those compartments. Thus, while comparing decomposition rates among temperate forests, small streams draining them, and peatlands and large lakes within the same catchment, one would predict fastest detrital decomposition in the lakes (low C:N:P owing to its planktonic origin), slower in forests and streams (high C:N:P in leaf litter), and slowest decomposition in peatlands (higher C:N:P and restricted decomposition due to anoxia). A higher proportion of detrital mass decomposition per unit time in aquatic ecosystems implies higher rates of nutrient recycling through the detrital compartment. Accordingly, turnover rates of nutrients through the detrital pool tend to be faster in aquatic than in terrestrial ecosystems.

•••

REFERENCES

Abdul-Wahab, A. and Rice, E.L. 1967. Plant inhibition by Johnson grass and its possible significance in oldfield succession. *Bull. Torrey Bot. Club* **94**: 486- 497.

Aber, J.D., McClaugherty, C.A and Melillo, J.M. 1984. Litter decomposition in Wisconsin forests-mass loss, organic chemical constituents and nitrogen. *School of Natural Resources, University of Wisconsin-Madison. Report* **R** 3284.

Aber, J.D. and Melillo, J.M. 1991. *Terrestrial Ecosystems*. Saunders College, Orlando, Fla.

Aber, J.D., Melillo, J.M. and McClaugherty, C.A. 1990. Predicting long term patterns of mass loss, nitrogen dynamics, and soil organic matter formation from initial fine litter chemistry in temperate forest ecosystems. *Can. J. Bot.* **68**: 2201-2208.

Ackerly, D.D. and Bazzaz, F.A. 1995. Leaf dynamics, self-shading and carbon gain in seedlings of a tropical pioneer tree. *Oecologia* 101: 289-298.

Adams, M.B. and Angradi, T.R. 1996. Decomposition and nutrient dynamics of hardwood leaf litter in the Fernow whole-watershed acidification experiment. *For. Ecol. Manage.* **83**: 61-69.

Aerts, R. 1990. Nutrient use efficiency in evergreen and deciduous species from heathlands. *Oecologia* **84**: 391-397.

Aerts, R. 1996. Nutrient resorption from senescing leaves of perennials: Are there general patterns? *J. Ecol.* **84**: 597-608.

Aerts, R. 1997 Climate, leaf litter chemistry and leaf litter decomposition in terrestrial ecosystem: A triangular relationship. *Oikos* **79**: 439-449.

Aerts, R., Bakker, C. and De Caluwe, H. 1992. Root turnover as a determinant of the cycling of C, N, and P in a dry heathland ecosystem. *Biogeochemistry* **15**:175-190.

Aerts, R. and Chapin, F.S.III. 2000. The mineral nutrition of wild plants revisited: a re-evaluation of processes and patterns. *Advances in Ecol. Res.* **30**: 1-67.

Aerts, R. and De Caluwe, H. 1997. Nutritional and plant-mediated controls on leaf litter decomposition of *Carex* species. *Ecology* **78**: 244-260.

Agren, GI., Kirschbaum, M.U.F., Johnson, D.W. and Bosatta, E. 1996. Ecosystem physiology-soil organic matter. In: A.L. Breymeyer, D.O. Hall, J.M. Melillo and G.I. Agra' (Eds.), *Global Change: Effects on Coniferous Forests and Grasslands*. Scope **56**. John Wiley & Sons, USA.

Ahlgren, I.F. 1974. The effect of fire on soil organisms. In: T.T. Kozlowski and C.E. Ahlgren (Eds.), *Fire and Ecosystems*, pp. 47-72. Academic Press, London.

Ahlgren, I.F. and Ahlgren, C.E. 1965. Effects of prescribed burning on soil microorganisms in Minnesota Jack pine forest. *Ecology* **46**: 306-310.

Ajtay, GL., Ketner, P. and Duvigneaud, P. 1979. Terrestrial primary producton and phytomass. In: B. Bolin, E.T. Degens, S. Kempe and P. Ketner, (Ed.), *The Global Carbon Cycle*, pp. 129-181. John Wiley and Sons, Chichester.

Ajwa, H.A., Dell, C.J. and Rice, C.W. 1999. Changes in enzyme activities and microbial biomass of tall grass prairie soil as related to burning and nitrogen fertilization. *Soil. Bio. Biochem.* **31**: 769-777.

Alexander, M. 1964. Biochemical ecology of soil microorganisms. *Annu. Rev. Microbiol.* **18**: 217-252.

Allan, J.D. 1995. *Stream ecology*. Chapman and Hall, London.

Amaranthus, M.P. and Trappe, J.M. 1993. Effects of erosion on ecto- and VA-mycorrhizal inoculum potential of soil following forest fire in south west Oregon. *Plant and Soil.* **150**: 41-49.

Amelung, W. and Zech, W. 1996. Organic species in ped surface and core fractions along a climosequence in the prairie, North America. *Geoderma* **74**: 193-206.

Amon, R.M.W. 2002. The biogeochemistry of dissolved organic matter in aquatic ecosystems—an overview. In: M. Bright, P.C. Dworschak and M. Stachowitsch (Eds.), *The Vienna School of Marine Biology: A tribute to Jörg Ott.*, pp. 1-18. Facultas Universitatsverlag, Wien.

Amthor, J.S. and Koch, G.W. 1996. Biota growth factor: Stimulation of terrestrial ecosystem net primary production by elevated atmospheric CO_2. In: G.W. Koch and H.A. Mooney (Eds.), *Carbon dioxide and Terrestrial Ecosystems*, pp. 399-414. Academic Press, New York.

Anderson, J.M. 1973a. The breakdown and decomposition of sweet chestnut (*Castanea sativa* Mill.) and beech (*Fagus sylvtica* L.) leaf litter in two deciduous woodland soils. I. Breakdown, leaching and decomposition. *Oecologia* **12**: 251-274.

Anderson, J.M. 1973b. The breakdown and decomposition of sweet chestnut (*Castanea sativa* Mill) and beech (*Fagus sylvatica* L.) leaf litter in two deciduous woodland soils. II. Changes in the carbon, nitrogen and polyphenol content. *Oecologia* **12**: 275-288.

Anderson, J.M. 1975. Succession, diversity and trophic relationships of some vsoil animals in decomposing leaf litter. *Journal of Animal Ecology* **44**: 475-495.

Anderson, J.M. 1992. Responses of soils to climate change. *Adv. Ecol Res.* **22**: 163-211.

Anderson, J.M. and Ineson, P. 1983. Interaction between soil arthropods andmicroorganisms in carbon, nitrogen and mineral element fluxes from decomposing leaf litter. In: J.A. Lee, S. Mc Neill, and I. H. Robinson (Eds.), *Nitrogen as an Ecological Factor*, pp. 413-432. Blackwell, Oxford

Anderson, J.M. and Meson, P. 1984. Interaction between microorganisms and soil invertebrates in nutrient flux pathways of forest ecosystems. In: J.M. Anderson, A.D. Rayner and D.W.H. Walton (Eds.), *Invertebrate Microbial Interactions*, pp. 59-88. Cambridge University Press, Cambridge.

Anderson, R.C. and Menges, E.S. 1997. Effects of fire on Sandhill herbs: nutrients, mycorrhizae and biomass allocation. *American Journal of Botany* **84**: 938-948.

Anderson, R.V., Coleman, D.C. and Cole, C.V. 1981. Effects of saprotrophic grazing on net mineralisation. In: F.E. Clark and T. Rosswall (Eds.), *Terrestrial Nitrogen Cycle: Processes, Ecosystem Strategies and Management Impacts*, pp. 201-216. Ecological Bulletins, Stockholm.

Andren, O., Bengtsson, J. and Clarhom, M. 1995. Biodiversity and species redundancy among litter decomposers. In: H.P. Collins, G.P. Robertson and M.J. Klug (Eds.), *The Significance and Regulation of Soil Biodiversity*, pp. 141-151. Kluwer Academic Publishers, Dordrecht, The Netherlands.

Andren, 0. and Paustian, K. 1987. Barley straw decomposition in the field: a comparison of models. *Ecology* **68**: 1190 - 1200.

Apinis, A.E. 1965. Thermophile Mikroorganismen in einegen dauergriinland-gesellschaften. *Biosoziologie* **10**: 290-303.

Armson, K.A. 1979. *Forest Soils: Properties and Processes*. University of Toronto Press, Toronto.

Arun Lekha. 1987. *Role of Soil Fauna in Soil Turnover and Litter Decomposition in an Agroforestry System.* M.Phil. Dissertation, Kurukshetra University, Kurukshetra, India.

Aston, A.R. and Gill, A.M. 1976. Coupled soil moisture, heat and water vapour transfer under simulated fire conditions. *Aust. J. Soil Res.* **14**: 55-56.

Atalla, R.H. 1999. Celluloses. In: B.M. Pinto (Ed.), *Comprehensive natural products chemistry, vol. 3. Carbohydrates and their derivatives including tannins. cellulose, and related lignin*, pp. 530-598. Elsevier, Amsterdam.

Attiwill, P.M. 1968. The loss of elements from decomposing litter. *Ecology* **49**: 142-145.

Azam, F. 1998. Microbial control of oceanic carbon flux: the plot thickens. *Science* **280**: 694-696.

Bakker, J.P. 1985. The impact of grazing on plant communities, plant populations and soil conditions on salt marshes. *Vegetatio* **62**: 391-398.

Baldock, J.A. and Skjemstad, J.O. 2000. Role of the soil matrix and minerals in protecting natural organic materials against biological attack. *Organic Geochemistry* **31**: 697-710.

Baldy, V. and Gessner, M.O. 1997. Towards a budget of leaf litter breakdown in a first order woodland stream. *C.R. Acad. Sci., Serie III, Sciences de la vie/ Life Sciences*. **320**: 747-758.

Ballare, C.L., Sanchez, R.A. Scopel, A.L. and Ghersa, C.M. 1988. Morphological responses of *Datura ferox* L. seedlings to the presence of neighbours. The relationship with canopy microclimate. *Oecologia* **76**: 288-293.

Barker, W.W., Welch, S.A. and Banfield, J.F. 1997. Biogeochemical Weathering of Silicate Minerals. In: J.F. Banfield and K.H. Nealson (Eds.), *Geomicrobiology: Interactions between Microbes and Minerals, Reviews in Mineralogy*, Vol. 35, pp. 81-122. Mineralogical Society of America, Washington, D.C.

Barlocher, F. 1982. Effects of drying and freezing autumn leaves on leaching and colonization by aquatic hypomycetes. *Freshwat. Biol.* **28**: 1-7.

Barlocher, F. 1985. The role of fungi in the nutrition of stream invertebrates. *J. Linn. Soc. Bot.* **91**: 83-94.

Barlocher, F. and Kendrick, B. 1974. Dynamics of the fungal populations on leaves in a stream. *J. Ecol.* **62**: 761-91.

Basaraba, J. and Starkey, R.L. 1966. Effects of plant tannins on decomposition of organic substances. *Soil Sci.* **101**: 17-23.

Beare, M.H., Parmelee, R.W., Hendrix, P.F., Cheng, W., Coleman, D.C. and Crossley Jr, D.A. 1992. Microbial and faunal interactions and effects on litter nitrogen and decomposition in agroecosystems. *Ecological Monographs* **62**: 569-591.

Beatty, S.W. and Sholes, O.D.V. 1988. Leaf litter effects on plant species composition of deciduous forest treefall pits. *Canad. J. Forest Res.* **18**: 553-559.

Becker, W.M. and Deamer, D.W. 1991. *The World of the Cell.* Benjamin/ Cummings, Redwood City, California.

Beerwerth, W. and Schurmarm, J. 1969. Contribution to the ecology of mycobacteria. *Zentbl. Bakt. Parasitkde, Abt.* I, **211**: 58-69.

Beldman, G, Mutter, M., Searle-van Leeuwen, M.J.F., van den Broek, L.A.M., Schols, H.A., aned Voragen, A.G.J. 1996. New enzymes active toward pectic structures. In: J. Visser and A.G.J. Voragens (Eds.), *Progress in Biotechnology 14-Pectins and Pectinases*, pp. 231-245. Elsevier, Amsterdam.

Bell, M.K. 1974. Decomposition of herbaceous litter. In: C.H. Dickinson and G.J.F. Pugh (Eds.), *Biology of plant litter decomposition*, vol. I, pp. 37-67. Academic Press, London and New York.

Benen, J.A.E., Kester, H.C.M., Parenicova, L., and Visser, J. 1996. Kinetics and mode of action of *Aspergillus niger* polygalacturonases. In: J. Visser and A.G.J. Voragens (Eds.), Progress in Biotechnology 14—*Pectins and pectinases*, pp. 221-230. Elsevier, Amsterdam.

Benfield, E.F. 1997. Comparison of litterfall input to streams. In: J.R. Webster and J.L. Meyer (Eds.), Stream organic matter budgets. *Journal of the North American Benthological Society*. **16**: 104-108.

Bengtsson, J., Zheng, D.W., Agren, G.I. and Persson, T. 1995. Food webs in soil: An interface between population and ecosystem ecology. In: C. Jones and J. Lawton (Eds.), *Linking Species and Ecosystems*, pp. 159-165. Chapman and Hall, New York.

Benke, A.C. and Wallace, J.B. 1990. Wood dynamics in coastal-plain blackwater streams. *Canadian Journal of Fisheries and Aquatic Sciences* **47**: 92-99.

Benthley, J.R. and Fenner, R.L. 1958. Soil temperatures during burning related to post fire seed beds on woodland range. *Journal of Forestry* **56**: 737-774.

Berendse, F. 1999. Implications of increased litter production for plant biodiversity. *Trends in Ecology and Evolution* **14**: 4-5.

Berg, B. 1984. Decomposition of root litter and some factors regulating the process: Long term root litter decomposition in a Scots pine forest. *Soil Biology and Biochemistry* **16**: 609-618.

Berg, B. 1986. Nutrient release from litter and humus in coniferous forest soils: a mini review. *Scandinavian Journal of Forest Research* **1**: 359-369.

Berg, B. and Agren, G.I. 1984. Decomposition of needle litter and its organic chemical components: Theory and field experiments. Long term decomposition in a Scots pine forest III. *Canadian Journal of Botany* **62**: 2880-2890.

Berg, B., Berg, M.P., Bottner, P., Box, E., Breymeyer, A., Calvo de Anta, R., Couteaux, M., Escudero, A., Gallardo, A., Kratz, W., Madeira, M., Mälkönen, E., McClaugherty, C., Meentemeyer, V., Mulioz, F., Piussi, P., Remade, J. and Virzo de Santo, A. 1993. Litter mass loss rates in pine forests of Europe and Eastern United States: some relationships with climate and litter quality. *Biogeochemistry* **20**: 127-159.

Berg, B., Booltink, H., Breymeyer, A., Ewertsson, A., Gallardo, A., Holm, B., Johansson, M.B., Koivuoja, S., Meentemeyer, V., Nyman, P., Olofsson, J., Reurslag, A., Staaf, H., Staaf, I. and Uba, L. 1991a. Data on needle litter decomposition and soil climate as well as site characteristics for some coniferous forest sites. Part 1. Site characteristics. *Dep. Ecol. Environ. Res., Swed. Univ. Agric. Sci. Rep.* **41**, Uppsala.

Berg, B., Booltink, H., Breymeyer, A., Ewertsson, A., Gallardo, A., Holm, B., Johansson, M. B., Koivuofa, S., Meentemeyer, V., Nyman, P., Olofsson, J., Reurslag, A., Staaf, H., Staaf, I. and Uba, L. 1991b. Data on needle litter decomposition and soil climate as well as site characteristics for some coniferous forest sites. Part 2. Decomposition data. *Dep. Ecol. Environ. Res., Swed. Univ. Agric. Sci. Rep*. **42**, Uppsala.

Berg, B., Calvo de Anta, R., Escudero, A., Gärdenäs, A., Johansson, M.B., Laskowski, R., Madeira, M., Mälkönen, E., McClaugherty, C., Meentemeyer, V. and Virzo de Santo, A. 1995a. The chemical composition of newly shed needle litter of Scots pine and some other pine species in a climatic transect. X. Long-term decomposition in a Scots pine forest. *Canadian Journal of Botany* **73**: 1423-1435.

Berg, B., McClaugherty, C., Virzo De Santo, A., Johansson, M.B., Ekbohm, 1995b. Decomposition of litter and soil organic matter- can we distinguish a mechanism for soil organic matter buildup? *Scandinavian Journal of Forest Research* **10**: 108-119.

Berg, B. and Ekbohm, G 1983. Nitrogen immobilization in decomposing needles at variable carbon:nitrogen ratios. *Ecology* **64**: 63-67.

Berg, B. and Ekbohm, G 1991. Litter mass loss rates and decomposition patterns in some needle and leaf litter types VII. Long term decomposition in a Scots pine forest. *Can. J. Bot.* **69**: 1449-1456.

Berg, B., Ekbohm, G., Johansson, M., McClaugherty, C., Rutigliano, F. and Virzo de Santo, A. 1996. Maximum decomposition limits of forest litter types: a synthesis. *Can. J. Bot.* **74**: 659-672.

Berg, B., Ekbohm, U and McClaugherty, C.A. 1984. Lignin and holocellulose relations during long term decomposition of some forest litters. Long-term decomposition in Scots pine forest. IV. *Can. J. Bot.* **62**: 2540-2550.

Berg, B., Ewertsson, A., Ferenczi, J., Holm, B., Jansson, P.E., Kuivuoja, S., Meentemeyer, V., Nyman, P., Staaf, H. and Staaf, I. 1986. Data on needle litter decomposition and soil climate as well as site characteristics for some coniferous forest sites. *Dep. Ecol. Environ. Res. Swed. Univ.Agric. Sci. Rep.* 22, Uppsala.

Berg, B. and Johansson, M.B. 1998. Maximum limit for foliar litter decomposition- a synthesis of data from forest systems. In: B. Berg (Ed.), *Maximum limits for plant litter decomposition, data from forest systems* pp. 4-100. Swedish University of Agricultural Sciences Report 77, Uppsala, Sweden.

Berg, B., Johansson, M.B., Meentemeyer, V. and Kratz, W. 1998. Decomposition of tree root litter in a climatic transect of coniferous forests in northern Europe: A synthesis. *Scandinavian Journal of Forest Research* **13**: 202-212.

Berg, B. and McClaugherty, C. 2003. *Plant litter. Decomposition, humus formation, carbon sequestration.* Springer-Verlag, Heidelberg.

Berg, B. and Meentemeyer, V. 2001. Litterfall in some European coniferous forests as dependent on climate—a synthesis. *Canadian Journal of Forest Research* **31**: 292-301.

Berg, M., Ruiter, P.D., Didden, W., Janssen, M., Schouten T. and Verhoef, H. 2001. Community food web, decomposition and nitrogen mineralization in a stratified Scots pine forest soil. *Oikos* **94**: 130-142.

Berg, B. and Staaf, H. 1980. Decomposition rate and chemical changes of Scots pine needle litter. II. Influence of chemical composition. In: T. Persson (Ed.), *Structure and Function of Northern Coniferous Forests: An Ecosystem Study*, pp. 373-390. Ecological Bulletins, Stockholm.

Berg, B., Staaf, H. and Wessen, B. 1987. Decomposition and nutrient release in needle litter from nitrogen-fertilized Scots pine *(Pinus sylvestris)* stands. *Scandinavian Journal of Forest Research* **2**: 399-415.

Berg, N.W. and Pawluk, S. 1984. Soil mesofaunal studies under different vegetative regimes in north Central Alberta. *Canadian Journal of Soil Science* **64**: 209-223.

Bhatta, B.K., Takeda, H. and Tokuchi, N. 2000. Effect of soil nitrogen on decomposition and nutrient dynamics of Japanese black pine litter. *Appl. For Sci. Kansai* **9**: 75-81.

Bilimoria, M.H. and Bhat, J.V. 1961. Microbial decomposition of pectic substances. *Part I.* Role of soil streptomycetes in the process *J. Sci. TechnoL,Cawnpore* **43**: 16-25.

Biondini, M., Klein, D.A. and Redente E.F. 1988. Carbon and nitrogen losses through root exudation by *Agropyron cristatum, A. smithii and Bouteloua gracilis. Soil Biol. Biochem.* **20**: 477-482.

Blair, J.M. 1988. Nutrient release from decomposing litter of three tree species with special reference to calcium, magnesium and potassium dynamics. *Plant and Soil* **110**: 49-55.

Bocock, K.L. 1964. Changes in the amounts of dry matter, nitrogen, carbon and energy in decomposing woodland leaf litter in relation to the activities of soil fauna. *J. Ecol.* **52**: 273-284.

Boer, W.D., Flman, L.B., Summerbell, R.C. and Boddy, L. 2005. Living in a fungal world: impact of fungi on soil bacterial niche development. *FEMS Microbiology Reviews* **29**: 795-811.

Boerner, R.E.J. 1984. Foliar nutrient dynamics and nitrogen use efficiency of four deciduous tree species in relation to site fertility. *J. Appl. Ecol.* **21**: 1029-1049.

Boerner, R.E.J. 1985. Foliar nutrient dynamics, growth and nutrient use efficiency of *Hamamelis virginiana* in three forest microsites. *Can. J. Bot.* **63**: 1476-1481.

Boiling, RH., Goodman, E.D., Van Sickle, J.A., Zimmer, J.0., Cummins, K.W., Petersen, R.C. and Reice, S.R. 1975. Towards a model of detritus processing in a woodland stream. *Ecology* **56**: 141-151.

Bokhari, V.G 1978. Allelopathy among prairie plants and its possible ecological significance. *Ann. Bot.* **42**: 127-136.

Boot, R.G.A. 1990. The significance of size and morphology of root systems for nutrient acquisition and competition. In: H. Lambers, M.L. Cambridge, H. Konings and T.L. Pons (Eds.), *Causes and Consequences of Variation in Growth Rate and Productivity of Higher Plants*, pp. 299-311. SPB, The Hague.

Bomebusch, C.H. 1930. The fauna of forest soil. *Forest. Fors. Vaes. Danm.* **11**: 1-224.

Box, E.O. 1975. Quantitative evaluation of global productivity models generated by computers. In: H.Leith and R.H.Whittaker (Eds.), *Primary Productivity of the Biosphere*, Ecological Studies Vol. 14. Springer-Verlag, New York.

Box, E.O. 1978. Geographical dimensions of terrestrial net and gross productivity. *Radiant. Environ. Biophys.* **15**: 305-322.

Bradford, M.A., Tordoff, G.M., Eggers, T., Jones, T.H. and Newington, J.N. 2002. Microbiota, fauna and mesh size interactions in litter decomposition. *Oikos* **99**: 317- 323.

Brady, N.C. and Weil, R.R. 2001. *The Nature and Properties of Soil*, 13th Edn. Prentice Hall, Englewood Cliffs, NJ.

Bray, J.R. and Gorham, E. 1964. Litter production in forests of the world. *Adv. Ecol. Res.* **2**: 101-187.

Bremer, D.J., Ham, J.M. and Owensby, C.E. 1996. Effects of elevated atmospheric carbon dioxide and open top chambers on transpiration in a tallgrass prairie. *J. Environ. Qual* **25**: 691-701.

Briand, F. 1983. Environmental control of food web structure. *Ecology* **64**: 253-263.

Brierley, J.K. 1955. Seasonal fluctuations in the oxygen and carbon dioxide concentrations in beech litter with reference to the salt uptake of beech mycorrhizas. *J. Ecol.* **43**: 404-408.

Briones, M.J.I. and Ineson, P. 1996. Decomposition of eucalyptus leaves in litter mixtures. *Soil Biol. Biochem.* 28: 1381-1388.

Broadfoot, W.M. and Pierre, W.H. 1939. Forests soil studies. I. Relation of rate of decomposition of tree leaves to their acid-base balance and other chemical properties. *Soil Sci.* **48**: 329-348.

Brock, T.D. 1969. In: P.M. Meadow and S.J. Pirt, (Eds.), *Microbial Growth*, pp. 15-41. Cambridge University Press.

Brock, T.C.M., De Lyon, M.J.H., Van Laar, E.M.J.M. and Van Loon, E.M.M. 1985. Field studies on the breakdown of *Nuphar lutea* (L.) SM. (Nymphaeaceae), and a comparison of three mathematical models for organic weight loss. *Aquat. Bot.* **21**: 1-22.

Brookes, P.C., Powlson, D.S., Jenkinson, D.S. 1984. Phosphorus in the soil microbial biomass. *Soil Biol. Biochem.* **16**: 169-175.

Bruederle, L.P and Stearns, F.W. 1985. Ice storms damage to a southern Wisconsin mesic forest. *Bull. Torrey Bot. Club* **112**: 167 - 175.

Bunnell, F.L., Maclean, S.F.Jr and Brown, J. 1975. Barrow, Alaska, USA. In: T. Rosswall and O.W. Heal (Eds.), *Structure and Function of Tundra Ecosystems*, pp. 73-124. Ecol. Bull. **20** Swedish Natural Science Research Council, Stockholm.

Bunnell, F.L. and Tait, D. 1974. Mathematical simulation models of decompositions process. In: A.J. Holding, O.W. Heal, S.F. Maclean Jr. and P.W. Flanagan (Eds.), *Soil Organisms and Decomposition in Tundra,* pp. 207-225. Tundra Biome Steering Committee, Stockholm.

Bunnell, F.L., Tait, D. and Flanagan, P.W. 1977a. Microbial respiration and substrate weight loss. II. A model of the influence of chemical composition. *Soil Biol. Biochem.* **9**: 41 -47.

Bunnell, F.L., Tait, D., Flanagan, P.W. and van Cleve, K. 1977b. Microbial respiration and substrate weight loss. I. A general model of the influence of abiotic variables. *Soil Biol. Biochem.* **9**: 33-40.

Cadish, G and Giller, K.E. 1997. *Driven by Nature. Plant Litter Quality and Decomposition.* CAB International, Wallingford.

Campbell, I.C. and Fuchshuber, L. 1994. Amount, composition and seasonality of terrestrial litter accession to an Australian cool temperate rain forest stream. *Arch. Hydrobiol.* **130**: 499-512.

Campbell, W.G. and Bryant, S.A. 1941. Determination of pH in wood. *Nature* **147**: 357.

Canadell, J.G, Pitelka, L.F. and Ingram, J.S.I. 1996. The effects of elevated [CO_2] on plant soil carbon below-ground: A summary and synthesis. *Plant and Soil* **163**: 121 - 130.

Cao, M. and Woodward, F.I. 1998. Net primary and ecosystem production and carbon stocks of terrestrial ecosystems and their responses to climate change. *Global Change Biology* **4**: 185 - 198.

Carlyle, J.C. 1986. Nitrogen cycling in forested ecosystems. *Forestry Abstracts* **47**: 307-336.

Carpenter, S.R. 1981. Decay of heterogeneous detritus: A general model. *J. Theor. Biol.* **89**: 539-547.

Carpenter, S.R. 1982. Comparisons of equations for decay of leaf litter in tree hole ecosystems. *Oikos* **39**: 17-22.

Carpenter, S.R. andAdams, M.S. 1979. Effects of nutrients and temperature on decomposition of Myriophyllum spicatum L. in a hard water eutrophic lake. *Limnol. Oceanogr* **24**: 520-528.

Carson, W.P. and Peterson, C.J. 1990. The role of litter in an old-field community: Impact of litter quantity in different seasons on plant species richness and abundance. *Oecologia* **85**: 8-13.

Carter, M.F and Grace, J.B. 1986. Relative effects of *Justicia americana* litter on germination, seedling and established plants of *Polygonum lapathifolium. Aquatic Bot.* **23**: 341 -349.

Casida, L.E. Jr. 1977. Microbial metabolic activity in soil as measured by dehydrogenase determinations. *Appl. Environ. Microbiol.* **34**: 630-636

Cebrian J. and Lartigue, J. 2004. Patterns of herbivory and decomposition in aquatic and terrestrial ecosystems. *Ecological Monographs* **74** :237 259.

Chamier, A.C. and Dixon, P.A. 1982. Pectinases in leaf degradation by aquatic hypomycetes: The enzymes and leaf maceration. *J. Gen. Microbiol.* **128**: 2469-2483.

Chapin, F.S.III. 1980. The mineral nutrition of wild plants. *Annu. Rev. Ecol. Syst.* **11**: 233-260.

Chapin, F.S.III. and Kedrowski, R.A. 1983. Seasonal changes in nitrogen and phosphorus fractions and autumn retranslocation in evergreen and deciduous taiga trees. *Ecology* **64**: 376-391.

Chapin, F.S.III., Matson, P.A. and Mooney, H.A. 2002. *Principles of Terrestrial Ecosystem Ecology*. Springer Verlag, USA.

Chapin, F.S.III. and Moilanen, L. 1991. Nutritional controls over nitrogen and phosphorus resorption from Alaskan birch leaves. *Ecology* **72**: 709-715

Chapin, F.S.III, van Cleve, K. and Chapin, M.C. 1979. Soil temperature and nutrient cycling in the tussock growth form of *Eriophorum vaginatum. J. Ecol.* **67**: 169-189.

Chen, Y. and Aviad, T. 1990. Effects of humic substances on plant growth. In: C.E. MacCarthy, R.L. Clapp, R.L. Malcolm and P.R. Bloom (Eds.), *Humic Substances in Soil and Crop Sciences: Selected Reading*, pp. 161-186. Soil Science Society of America, Madison, Wisconsin.

Chen, Y., Clapp, C.E., Magen, H. and Cline, V.W. 1999. Stimulation of Plant Growth by Humic Substances: Effects on Iron Availability. In: E.A. Ghabbour and G. Davies (Eds.), *Understanding Humic Substances: Advanced Methods, Properties and Applications,* pp. 255-263. Royal Society of Chemistry, Cambridge, UK.

Chen, Y., Magen, H., and Clapp, C.E. 2001. Plant Growth Stimulation by Humic Substances and Their Complexes with Iron. *Proceedings of the International Fertilizer Society Symposium*, Lisbon, March 2001.

Chertov, O.G. and Komarov, A.S. 1997. SOMM: A model of soil organic matter dynamics. *Ecological Modelling* **94**: 177-189.

Choudhury, D. 1988. Herbivore induced changes in leaf litter resource quality: A neglected aspect of herbivory in ecosystem nutrient dynamics. *Oikos* **51**: 389-393.

Christensen, 0. 1975. Wood litter fall in relation to abscission, environmental factors, and the decomposition cycle in a Danish oak forest. *Oikos* **26**: 187-195.

Claeys, M., Graham, B., Vas, G., Wang, W., Vermeylen, R., Pashynska, V.C., Guyon, P., Andreae, M.O., Artaxo, P. and Maenhaut, W. 2004. Formation of secondary organic aerosols through photooxidation of isoprene. *Science* **303**: 1173-1176.

Clapp, C.E. 2001. An Organic Matter Trail: Polysaccharides to Waste Management to Nitrogen/Carbon to Humic Substances. In: E.A. Ghabbour and G. Davies (Eds.), *Humic Substances: Structures, Models and Functions*, pp. 3-17. Royal Society of Chemistry, Cambridge, UK.

Clapp, C.E., Chen, Y., Hayes, M.H.B., and Cheng, H.H. 2001. Plant Growth Promoting Activity of Humic Substances. In: R.S. Swift and K.M. Spark (Eds.), *Understanding Organic matter in Soils, Sediments and Waters.* Proceedings of the 9th International Conference of the International Humic Substances Society, Adelaide, Australia, 21st-25th September 1998, IHSS St Paul, Minnesota.

Clein, J.S., Kwiatkowski, B.L., McGuire, A.D., Hobbie, J.E., Rastetter, E.B., Melillo, J.M. and Kicklighter, D.W. 2000. Modeling carbon responses of tundra ecosystems to historical and projected climate: a comparison of a plot and a global scale ecosystem model to identify process based uncertainties. *Global Change Biology* **6**: 127-140.

Clien, J.S. and Schimel, J.P. 1994. Reduction in microbial activity in birch litter due to drying and rewetting events. *Soil Biology and Biochemistry* **26**: 403-406.

Coelho Neto, A.L. 1987. Overland flow production in a tropical rain forest catchment: The role of litter cover. *Catena* 14: 213-232.

Coffin, D.P and Lauenroth, W.K. 1996. Transient responses of North -American grasslands to changes in climate. *Climatic Change* **34**: 269-278.

Cole, J.J. and Caraco, N.F. 2001. Carbon in catchments: connecting terrestrial carbon losses with aquatic metabolism. *Mar. Freshwater Res.* **52**: 101-110.

Cole, J.J., Pace, M.L. Carpenter, S.R. and Kitchell, J.F. 2000. Persistence of net heterotrophy in lakes during nutrient addition and food web manipulations. *Limnol. Oceanogr.* **45**: 1718-1730.

Coleman, D.C. 1994. The microbial loop concept as used in terrestrial soil ecology studies. *Microbial Ecology* **28**: 245-250.

Coley, P.D., Bryant, J.P. and Chapin, F.S.III. 1985. Resource availability and plant anti-herbivore defense. *Science* **230**: 895-899.

Collins, B.S and Quinn, J.A. 1982. Displacement of *Andropogon scoparius* on the New Jersey Piedmont by the successional shrub *Myrica pennsylvanica. Amer J. Bot.* **69**: 680-689.

Collins, S.L. and Gibson, D.J. 1990. Effects of fire on community structure in tallgrass and mixed grass prairie. In: S.L. Collins and L.L. Wallace (Eds.), *Fire in North American Tallgrass Prairies,* pp. 81-98. University of Oklahema Press, Norman.

Collins, S.L. and Good, R.E. 1987. The seedling regeneration niche: Habitat structure of tree seedlings in an oak- pine forest. *Oikos* **48**: 89-98.

Conn, C. and Dighton, J. 2000. Litter quality influences on decomposition, ectomycorrhizal community structure and mycorrhizal root surface acid phosphatase activity. *Soil Biol. Biochem.* **32**: 489-496.

Corbeels, M. 2001. Plant litter and decomposition. General concepts and model approaches. *NEE workshop proceedings 18th-20th April,* pp. 124-129

Corke, C.T. and Chase, F.E. 1964. Comparative studies of actinomycete populations in acid podzolic and neutral mull forest soils. *Proc. Soil Sci Soc. Am.* **28**: 68-70.

Cornelissen, J.H.C. 1996. An experimental comparison of leaf decomposition rates in a wide range of temperate plant species and types. *Journal of Ecology* **84**: 573-582.

Cortez, J. and Bouche, M.B. 1998. Field decomposition of leaf litters: Earthworm - microorganisms interactions-the ploughing-in effect. *Soil Biology and Biochemistry* **30**: 795-804.

Cortez, J. and Hameed, R. 1988. Effects de la maturation des litieres de ray-grass (*Lolium perenne* L.) dans le sol sur leur consommation et leur assimilation par *Lumbricus terrestris L. Revue Ecologie et Biologie du Sol* **25**: 397-412.

Côté, B., Fyles, J. W. and Djalilvand, H. 2002. Increasing N and P resorption efficiency and proficiency in northern deciduous hardwoods with decreasing foliar N and P concentrations. *Annals of Forest Science* **59**: 275-281.

Cotrufo, M.F. and Ineson, P. 1996. Elevated CO_2 reduces field decomposition rates of *Betula pendtda* Roth. leaf litter. *Oecologia* **106**: 525-530.

Cotrufo, M.F., Berg, B. and Kratz, W. 1998a. Increased atmospheric CO_2 and litter quality. *Environ. Rev.* **6**: 1-12.

Cotrufo, M.F., Ineson, P. and Scott, A. 1998b. Elevated CO 2 reduces the nitrogen concentration of plant tissues. *Global Change Biology* **4**: 53-54.

Cotrufo, M.F., Briones, M.J.I. and Meson, P. 1998c. Elevated CO_2 effects, field decomposition rate and palatability of tree leaf litter: Importance of changes in substrate quality. *Soil Biology Biochemistry* **30**: 1565-1571

Cotrufo, M.F., Meson, P. and Rowland, A.P. 1994. Decomposition of tree leaf litters grown under elevated CO_2 : effect of litter quality. *Plant Soil* **163**: 121-130.

Cotrufo, M.F., Ineson, P. and Roberts, J.D. 1995. Decomposition of birch leaf litter with varying C to N ratios. *Soil Biol. Biochem.* **27**: 1219-1221.

Couteaux, M.M., Bottner, P. and Berg, B. 1995. Litter decomposition, climate and litter quality. *Tree* **10**: 63-66.

Couteaux, M.M., McTiernan, K.B., Berg, B., Szuberla, D., Dardenne, P. and Bottner, P. 1998. Chemical composition and carbon mineralization potential of Scots pine needles at different stages of decomposition. *Soil Biol. Biochem.* **30**: 583-595.

Covich, A.P. 1988. Geographical and historical comparisons of neotropical streams: biotic diversity and detrital processing in high variable habitats. *J. N. Am. Benthol.* Soc. **7**: 361-386.

Covich, A.P. and Crowl, T.A.1990. Effects of hurricane storm flow on transport of woody debris in a rain forest stream (Luquillo Experimental Forest, Puerto Rico). In: J.H. Krishna, V. Quinones-Aponte, F. Gomez and G.L. Morris (Eds.), *Tropical Hydrology and Caribbean Water Resources,* pp. 197-205. American Water Resources Association, Bethesda, MD.

Crestini, C., Bernini, R., Porri, A., and Giovannozzi-Sermanni, G. 1996. Biodegradation of monomeric, dimeric and polymeric lignin models by Lentinus edodes. *Holzforschung* **50**: 193-200.

Crestini, C., Sermanni, G.G., and Argyropoulos, D. S. 1998. Structural modifications induced during biodegradation of wheat lignin by *Lentinula edodes. Bioorganic and Medicinal Chemistry* **6**: 967-973.

Cuffney, T.F., Wallace, J.B. and Lugthart, G.J. 1990. Experimental evidence quantifying the role of benthic invertebrates in organic matter dynamics of headwater streams. *Freshwater Biology* **23**: 281-299.

Cummins, K.W. 1974. Structure and function of stream ecosystems. *Bioscience* **24**:61-64.

Cummins, K.W., Wilzbach, M.A., Gates, D.M., Perry, J.B. and Taliaferro, W.B. 1989. Shredders and riparian vegetation. *Bioscience* **39**: 24-30.

Curtis, P.S. 1996. A meta-analysis of leaf gas exchange and nitrogen in trees grown under elevated carbon dioxide. *Plant Cell Environ.* **19**: 127-137.

Curtis, P.S. and Wang, X. 1998. A meta-analysis of elevated CO 2 effects on woody plant mass, form and physiology. *Oecologia* **113**: 299-313.

Cushing, C.E., Minshall, G.W. and Newbold, J.D. 1993. Transport dynamics of fine particulate organic matter in two Idaho streams. Limnology and *Oceanography* **38**: 1101-1115.

Dagley, S. 1967. The microbial metabolism of phenolics. In: A.D. McLaren and G.H. Paterson (Eds.), *Soil Biochemistry*, pp. 287-317. Marcel Dekker, New York.

Dalal, R.C. 1975. Urease activity in some Trinidad soils. *Soil Biol. Biochem.* **7**: 5-8.

Dalal, R.C. 1979. Mineralization of carbon and phosphorus from carbon-14 and phosphorus-32 labeled plant material to soil. *Soil Sci.Soc. Am. J.* **43**: 913-916.

Dalal, R.C. 1998. Soil microbial biomass-what do the numbers really mean? *Aust. J. Exp. Agric.* **38**: 645-665.

Dalal, R.C. 2001. Acidic soil pH, aluminum and iron affect organic carbon turnover in soil, pp. 111-114. *NEE workshop Proceedings* 18th *-* 20th *April..*

Dar, G.H. 1995. Effect of heavy metals (Cd, Cr, Ni and Pb) on soil microbial biomass, carbon minerialization and enzyme activities. *International Journal of Ecology and Environmental Sciences* **21**: 87-95.

David, J.F., Malet, N., Couteaux, M.M. and Roy. J. 2001. Feeding rates of the woodlouse *Armadillidium vulgare* on herb litters produced at two levels of atmospheric CO_2 . *Oecologia* 127: 343-349.

Davies, F.L. and Williams, S.T. 1970. Studies. on the ecology of actinomycetes in soil. I. The occurrence and distribution of actinomycetes in a pine forest. *Soil Biol. Biochem.* **2**: 227- 238.

Day, F.P. 1983. Effects of flooding on leaf litter decomposition in microcosms. *Oecologia* **56**: 180-184.

Day, K.S., Thornton, R. and Kreeft, H., 2000. Humic acid products for improved phosphorus fertilizer management. In: E.A. Ghabbour, (Ed.), *Humic Substances, Versatile Components of Plants, Soil and Water*, pp. 321-325. Royal Society of Chemistry, London, UK.

De Jong, T.J. and Klinkhamer, P.G.L. 1985. The negative effect of litter ofparent plants of *Cirsium vulgare* to their offsprings. Autotoxicity or immobilization. *Oecologia* **65**: 153-166.

Deregibus, V.A., Sanchez, R.A, Casal, J.J. and Trlica, M.J. 1985. Tillering responses to enrichment of red light beneath the canopy in humid natural grasslands. *J. Appl. Ecol.* **22**: 199-206.

Dhillon, S.S. and Anderson, R.C. 1993. Growth dynamics and associated mycorrhizal fungi of little bluestem grass (*Schizachyrium scuparium* Nash.) on burned and unburned sand prairies. *New Phytologist* **123**: 77-91.

Diaz, S., Grime, J.P., Harris, J. and Mcpherson, E. 1993. Evidence of a feedback mechanism limiting plant response to elevated carbon dioxide. *Nature* **364**: 616-617.

Dick., R.P. 1994. Soil enzyme activities as indicators of soil quality. In: J.W. Doran, D.C. Coleman,D.F. Bezdicek, and B.A. Stewart (Eds.), *Defining Soil Quality for a Sustainable Environment*, pp. 107-124. American Society of Agronomy, Madison. WI.

Dick, R.P., Myrold, D.D. and Kerle, E.A. 1988a. Microbial biomass and soil enzyme activities in compacted and rehabilitated skid trail soils. *Soil Science Society of America Journal* **52**: 512-516.

Dick, R.P., Rasmussen, P.E. and Kerle, E.A. 1988 b. Influence of long term residue management on soil enzyme activity in relation to soil chemical properties of a wheat fallow system. *Biology and Fertility of Soils* **6**: 159-164.

Dick, W.A and Tabatabai, M.A. 1992. Potential uses of soil enzymes. In: F.B. Metting Jr. (Ed.), *Soil Microbial Ecology: Applications in Agricultural and Environmental Management*, pp. 95-127. Marcel Dekker, New York, USA

Dickinson, C.H. and Pugh, G.J.F. 1974. *Biology of Plant Litter Decomposition* Vol. II. Academic Press, London and NewYork.

Dilly, 0. and Munch, J.C. 1996. Microbial biomass content, basal respiration and enzyme activities during the course of decomposition of leaf litter in a black alder (*Alnus glutinosa* (L.) Gaertn.) forest. *Soil Biol. Biochem.* **28**: 1073-1081.

Dix, R.L. 1960. The effect of burning on the mulch structure and species composition in grasslands in westerns south Dakota. *Ecology* **41**: 49-56.

Doetsch, R.N. and Cooke, T.M. 1973. *Introduction to Bacteria and Their Ecobiology*. Medical and Technical Publishing Co. Ltd. Lancaster.

Dormaar, J.F., Johnston, A. and Smoliak, S. 1984. Seasonal changes in carbon content and dehydrogenase, phosphatase and urease activities in mixed prairies and fescue grassland Ah horizons. *J. Range Manage.* **37**: 31-35.

Dormaar, J.F., Smoliak, S. and Willms, W.D. 1989. Vegetation and soil responses to short duration grazing on fescue grasslands. *J. Range Manage.* **42**: 252-256.

Douglas, C.J. 1996. Phenylpropanoid metabolism and lignin biosynthesis, from weeds to trees. *Trends in Plant Science* **1**:17 1-17 8 .

Dubey, R.C. and Maheshwari, D.K. 1999. *A Text Book of Microbiology*. S. Chand and Company, New Delhi, India.

Dukes, J.S. and Hungate, B.A. 2002. Elevated carbon dioxide and litter decomposition in California annual grasslands: Which mechanisms matter? *Ecosystems* **5**: 171-183.

Dyksterhuis, E.J. and Schmutz, E.M. 1947. Natural mulches or "litter" of grasslands, with kinds and amounts on a southern prairie. *Ecology* **28**: 163-179.

Eckstein, R.L. and Donath, T.W. 2005. Interactions between litter and water availability affect seedling emergence in four familial pairs of floodplain species. *J. Ecol.* **93**: 807-816.

Eckstein, R.L., Karlsson, P.S. and Weih, M. 1999. Leaf life span and nutrient resorption as determinants of plant nutrient conservation in temperate-arctic regions. *New Phytologist* **143**: 177-189.

Edmonds, R.L. 1984. Long-term decomposition and nutrient dynamics in Pacific silver fir needles in western Washington. *Can. J. For. Res.* **14**: 395-400.

Edwards, C. A. and Lofty, J. R. 1977. *Biology of Earthworms* (2nd edition). Chapman & Hall, London.

Ehmian, T.P. and Lamberti, GA.1992. Hydraulic and particulate matter retention in a 3rd-order Indiana stream. *Journal of the North American Benthological Society*. **11**: 341-349.

Ellenberg, H. 1988. *Vegetation Ecology of Central Europe*. Cambridge University Press, Cambridge.

Ellen, B.H. and Bettany, J.R. 1992. Temperature dependence of net nitrogen and sulfur mineralization. *Soil Sci. Soc. Am. J.* **56**: 1133-1141.

Elliott, J.M. 1971. The distances traveled by drifting invertebrates in a Lake District stream. *Oecologia* **6**: 350-379.

Enriquez, S., Durate, C.M. and Sand-Jensen, K. 1993. Patterns in decomposition rates among photosynthetic organisms: the importance of detritus C: N: P content. *Oecologia* **94**: 457-471.

Evans, R.A. 1972. Germination and establishment of Salsola in relation to seedbed environment II. Seed distribution, germination and seedling growth of Salsola and micro-environment monitoring of the seedbed. *Agric. J.* **64**: 219-224.

Evans, R.A. and Young, J. 1970. Plant litter and establishment of alien annual weed species in rangeland communities. *Weed Science* **18**: 697-703.

Everett, R.A. and Ruiz GM. 1993. Coarse woody debris as a refuge from predation in aquatic communities - an experimental test. *Oecologia* **93**: 475-486.

Facelli, J.M. 1988. Response to grazing after nine years of cattle exclusion in a flooding Pampa grassland, Argentina. *Vegetatio* **78**: 21-25.

Facelli, J.M., D'Angela, E. and Leon, R.J.C. 1987. Diversity changes during pioneer stages in a sub-humid pampean grassland succession. *Amer. Midl. Nat.* **117**: 17-25.

Facelli, J.M., Montero, C.M. and Leon, R.J.C. 1988. Effect of different disturbance regimen on semi-natural grasslands from the sub-humid Pampa. *Flora* **180**: 241-249.

Facelli, J.M. and Pickett, S.T.A. 1991 a. Plant litter: Its dynamics and effects on plant community structure. *Botanical Review.* **57**: 1-32.

Facelli, J.M and Pickett, S.T.A 1991b. Plant litter: Light interception and effects on an old field plant community. *Ecology* **72**: 1024-1031.

Fausch, K.D. and Northcote T.G. 1992. Large woody debris and Salmonid habitat in a small coastal British Columbia stream. *Canadian Journal of Fisheries and Aquatic Sciences* **49**: 682-693.

Feild, T.S., Lee, D.W. and Holbrook, N.M. 2001. Why leaves turn red in autumn? The role of anthocyanins in senescing leaves of red-osier dogwood. *Plant Physiology* **127**: 566-574.

Feller, U. and Fischer, A. 1994. Nitrogen metabolism in senescing leaves. *Critical Reviews in Plant Sciences* **13**: 241-273.

Fenn, L.B., Hasanein, B. and Burks, C.M. 1995. Calcium-ammonium effects on growth and yield of small grains. *Agronomy Journal* **87**: 1041-1046.

Fenner, M. 1985. *Seed Ecology*. Chapman & Hall, New York.

Fergus, C.L. 1969. The cellulolytic activity of thermophilic fungi and actinomycetes. *Mycologia* **61**: 120-129.

Findlay, S.E.G. and Arsuffi, T.L. 1989. Microbial growth and detritus transformations during breakdown of leaf litter in a stream. *Freshwat. Biol.* **21**:-261-269.

Fisher, S.G. and Likens, GE. 1973. Energy flow in Bear Brook, New Hampshire: An integrative approach to stream ecosystem metabolism. *Ecol. Monogr.* **43**: 421-39.

Flanagan, P.W. and van Cleve, K. 1983. Nutrient cycling in relation to composition and organic matter quality in taiga ecosystems. *Canad. J. Forest Res.* **13**: 795-817.

Fog, K. 1988. The effect of added nitrogen on the rate of decomposition of organic matter. *Biological Review* **63**: 433-462.

Fogel, R. 1980. Mycorrhizae and nutrient cycling in natural forest ecosystems. *New Phytologist* **86**: 199-212.

Fogel, R. and Cromack, K.Jr. 1977. Effect of habitat and substrate quality on Douglas fir litter decomposition in western Oregon. *Canadian Journal of Botany* **55**: 1632-1640.

Fogel, R. and Hunt, G. 1979. Fungal and arboreal biomass in a western Oregon Douglas-fir ecosystem: distribution patterns and turnover. *Canadian Journal of Forest Research 9*: 245-256.

Fogel, R. and Hunt G. 1983. Contribution of mycorrhizae and soil fungi to nutrient cycling in a Douglas-Fir ecosystem. *Canadian Journal of Forest Research* **13**:219-232.

Force, E.G. and McCarty, P.L. 1970. Anaerobic decomposition of algae. *Environ. Sci. Tech.* **4**: 842-49.

Foster, B.L. and Gross, K.L. 1998. Species richness in a successional grassland: Effect of nitrogen enrichment and plant litter. *Ecology* **79**: 2593-2602.

Fowler, N.L. 1986. Microsite requirement for germination and establishment of three grass species. *Amer. Midl. Nat.* **115**: 131-145.

Franck, V.M., Hungate, B.A., Chapin, F.S.III and Field, C.B. 1997 Decomposition of litter produced under elevated CO_2 : Dependence on plant species and nutrient supply. *Biogeochemistry* **36**: 223-237.

Frank, T. and Malkomes, H.P. 1993. Influence of temperature on microbial activities and their reaction to the herbicide Goltix in different soils under laboratory conditions. *Zentralblatt fur Mikrobiologie* **148**: 403-412.

Frankenberger, W.T.Jr. and Dick, W.A. 1983. Relationships between enzyme activities and microbial growth and activity indices in soil. *Soil Science Society of American Journal* **47**: 945-951.

Frankland, J.C. 1966. Succession of fungi on decaying petioles of *Pteridium aquilinum. J. Ecol.* **54**: 41-63.

Frankland, J.C. 1969. Fungal decomposition of bracken petioles. *J. Ecol.* **57**: 25-36.

Frankland, J.C. 1976. Decomposition of bracken litter. Bot. *J. Linn. Soc.* **73**: 133-143.

Frankland, J.C., Ovington, J.D. and Macrae, C. 1963. Spatial and seasonal variation in soil, litter and ground vegetation in some Lake District Woodlands. *J. Ecol.* **51**: 97-112.

Friesen, D.K., and Blair, G.M. 1988. A dual radio tracer study of transformations of organic, inorganic and plant residue phosphorus in soil in the presence and absence of plants. *Aust. J. Agric. Res.* **26**: 355-366.

Froment, A. 1972. Soil respiration in a mixed oak forest. *Oikos* **23**: 273-277.

Furniss, P.R. and Ferrar, P. 1982. A model of Savanna litter decomposition. *Ecol. Model.* **17**: 33-51.

Gadgil, R.L. and Gadgil, P.D. 1971. Mycorrhiza and litter decomposition. *Nature* **233**: 133.

Gadgil, R.L. and Gadgil, P.D. 1975. Suppression of litter decomposition by mycorrhizal roots of Pinus radiata. *New Zealand Journal of Forest Science* **5**: 33-41.

Gahrooee, F.R. 1998. Impacts of elevated atmospheric CO_2 on litter quality, litter decomposability and nitrogen turnover rate of two oak species in a Mediterranean forest ecosystem. *Global Change Biology* **4**: 667-677.

Garca, M.A.S. 1993. Patterns and processes in detritus based stream systems. *Limnologica* **23**: 107-114.

Garnett, E., Jonsson, L.M., Dighton, J. and Murnen, K. 2004. Control of pitch vine seed germiantion and initial growth exerted by leaf litters and polyphenolic compounds. *Biol. Fertil. Soils.* **40**: 421-426.

Garrett, S.D. 1981. *Soil Fungi and Soil Fertility.* 2nd Ed Pergamon Press, Oxford

Gartner, T.B. and Cardon, Z.G. 2004. Decomposition dynamics in mixed-species leaf litter. *Oikos* **104**: 230-246.

Gessner, M.O. 1999. A perspective on leaf litter breakdown in streams. *Oikos* **85**: 377-384.

Gessner, M.O. and Chauvet, E. 1994. Importance of stream microfungi in controlling breakdown rates of leaf litter. *Ecology* **75**: 1807-1817.

Ghabbour E.A. and Davies G. 2001. *Humic Substances: Structures, Models and Functions.* Royal Society of Chemistry, Cambridge.

Gholz, H.L., Wedin, D.A., Smitherman, S.M., Harmon, M.E. and Parton, W.J. 2000. Long term dynamics of pine and hardwood litter in contrasting environments: Towards a global model of decomposition *Global Change Biology* **6**: 751-765.

Giardina, C.P. and Ryan, M.G. 2000. Evidence that decomposition rates of organic carbon in mineral soil do not vary with temperature. *Nature* **404**: 858-861.

Gibson, D.J. and Hetrick, B.A.D. 1988. Topographic and fire effects on the composition and abundance of VA-mycorrhizal fungi in tallgrass prairie. *Mycologia* **80**: 433-441.

Gibson, D.J. and Hulbert, L.C. 1987. Effects of fire, topography and year to year climatic variations on species composition in tallgrass prairie. *Vegetation* **72**: 175-185.

Gijsman, A.J, Alarcon, H.F. and Thomas, R.J. 1997. Root decomposition in tropical grasses and legumes as affected by soil texture and season. *Soil Biol. Biochem.* **29**: 1443-1450.

Glinski, J. and Lipiec, J. 1990. *Soil Physical Conditions and Plant Roots*. CRC Press, Boca Raton, FL.

Golley, F.B. 1965. Structure and function of an oldfield broomsedge community. *Ecol. Monogr.* 35: 113-137.

Golley, F.B and Gentry, J.B. 1966. A comparison of variety and standing crop of vegetation on a one-year and twelve year abandoned field. *Oikos* **15**: 185-199.

Goncalves, J.L.M. and Carlyle, J.C. 1994. Modeling the influence of moisture and temperature on net nitrogen minerialization in a forested sandy soil. *Soil Biol. Biochem.* **26**: 1557-1564.

Goodfellow, M, 1971. Numerical taxonomy of some nocardioform bacteria. *J. Gen. Microbiol.* **69**: 33-80.

Goodfellow, M. and Cross, T. 1974. Actinomycetes. In: C.H. Dickinson and G.J.F. Pugh (Eds.), *Biology of Plant Litter Decomposition*, Vol. 2., pp. 269-302. Academic Press, London and New York.

Goodfellow, M., Hill, I.R. and Gray, T.R.G. 1968. Bacteria in a pine forest soil. In: T.R.G. Gray and D. Parkinson (Eds.), *The Ecology of Soil Bacteria*, pp. 500-514. Liverpool University Press.

Gosz, J.M., Likens, G.E. and Bormman, F.H. 1973. Nutrient release from decomposing leaf and branch litter in the Hubbard Brook forest, New Hempshire. *Ecological Monographs* **43**: 173-191.

Graca, M.A.S. 1993. Patterns and process in detritus based stream systems. *Limnologica* **23**: 107-114.

Gray, T.R.G. and Baxby, P. 1968. Chitin decomposition in soil. II. The ecology of chitinoclastic microorganisms in forest soil *Trans. Br. Mycol. Soc.* **51**: 293-309.

Gray, T.R.G. and Williams, S.T. 1971. Microbial productivity in soil. In: D. Hughes and A.H. Rose (Eds.), *Microbes and Biological Productivity,* pp. 255-286. Symp. Soc. Gen. Microbiol. Cambridge University Press, London.

Greenwood, D.J. 1968. Measurement of microbial metabolism in soil. In: T.R.G. Gray and D. Parkinson (Eds.), *Ecology of Soil Bacteria*, pp. 138-151. Liverpool University Press, UK.

Griffin, D.M. 1963. Soil moisture and the *ecology of soil fungi. Biol. Rev.* **38**: 141-166.

Griffm, D.M. 1972. Ecology of Soil Fungi. Chapman and Hall, London.

Griffiths, B.S. 1994. Soil nutrient flow. In: J. Darbyshire (Ed.), *Soil Protozoa*, pp. 65-91. CAB International, Wallingford, Oxon, UK.

Grime, J.P. 1979. *Plant Strategies and Vegetation Processes*. John Wiley and Sons, New York.

Grimm, N.B., Gergel, S.E., McDowell, W.H., Boyer, E.W., Dent, C.L., Groffinan, P., Hart, S.C., Harvey, J., Johnston, C., Mayorga, E., McClain, M.E. and Pinay, G. 2003. Merging aquatic and terrestrial perspectives of nutrient biogeochemistry. *Oecologia* **442**: 485-501.

Gross, GG. 1999. Biosynthesis of hydrolysable tannins. In: B.M. Pinto (Ed.), *Comprehensive Natural Products Chemistry, Carbohydrates and Their Derivatives Including Tannins, Cellulose, and Related Lignin.* Vol. 3, pp. 799-826. Elsevier, Amsterdam.

Grubb, P.J. 1977. The maintenance of species richness in plant communities: The importance of the regeneration niche. *Biol. Rev.* **82**: 107-145.

Gulledge, J. Doyle, A. and Schimel, J. 1997. Different NH_4^+ inhibition patterns of soil CH_4 consumption: A result of distinct CH_4 oxidizer populations across sites? *Soil Biol. Biochem.* **29**: 13-21.

Gunnarsson, T., Syndin, P. and Tunlid, A. 1988. The importance of leaf litter fragmentation for bacterial growth. *Oikos* **52**: 303-308.

Gupta. S.R. and Arun Lekha 1989. Decomposition and carbon turnover in grassland, forest and agricultural systems in a dry - subhumid region. In: J.S. Singh and Brij Gopal (Eds.), *Perspectives in Ecology*, pp. 131-168. Jagmander Book Agency, New Delhi.

Gupta, S.R. and Singh, J.S. 1977. Decomposition of litter in a tropical grassland. *Pedobiologia* **17**: 330-333.

Gupta, S.R. and Singh, J.S. 1981 a. The effect of plant species, weather variables and chemical composition of plant material on decomposition in a tropical grassland. *Plant and Soil* 59: 99-117.

Gupta, S.R. and Singh, J.S. 198 lb. Siol respiration in a tropical grassland. *Soil. Biol. Biochem*. **13**: 261-268.

Guzman, L. 1997. Contributions of decomposition to the carbon cycle. *Ecology Term Paper*, Fall 1997.

Gyllenberg, H.G. and Eklund, E. 1974. Bacteria. In: C.H. Dickinson and G.J.F. Pugh (Eds.). *Biology of Plant Litter Decomposition* Vol. 2, pp. 245-268. Academic Press London, New York.

Hagvar, S. 1988. Decomposition studies in an easily constructed microcosm: Effect of microarthropods and varying soil pH. *Pedobiologia* **31**: 293-303.

Haider, K., Frederick, L.R. and Flaig, W. 1965. Reactions between amino-acid compounds and phenols during oxidation. *Plant and Soil* **22**: 49-64.

Hairston, N.G. Jr. and Hairston, N.G. Sr. 1993. Cause effect relationships in energy flow, trophic structure, and interspecific interactions. *Am. Nat.* **142**: 379-411.

Hall, R.O., Wallace J.B. and Eggert S.L. 2000. Organic matter flow in stream food webs with reduced detrital resource base. *Ecology* **81**: 3445-3463.

Hamrick, J.L. and Lee, J.M. 1987. Effects of soil surface topography and litter cover on germination, survival and growth of musk thistle. *Amer. J. Bot.* **74**: 451-457.

Hansen, R.A. and Coleman, D.C. 1998. Litter complexity and composition are determinants of the diversity and species composition of orabatid mites (Acari: Oribatida) in litterbags. *Appl. Soil Ecol*. **9**: 17-23.

Hanson, B.J., Cummins, K.W., Barnes, J.R., Carter, M.W. 1984. Leaf litter processing in aquatic systems: A two variable model. *Hydrobiologia* **111**: 21-31.

Harley, J.L. 1971. Fungi in ecosystems. *J. Ecol*. **59**: 653-668.

Harmon, M.E. and Chen, H. 1991. Coarse woody debris dynamics in two old-growth ecosystems: comparing a deciduous forest in China and conifer forest in Oregon. *BioScience* **41**: 604-610.

Harmon, M.E., Franklin, J.F., Swanson, F.J. Sollins, P., Gregory, S.V., Lattin, J.D., Anderson, N.H. Cline, S.P. Aumen, N.G., Sedell, J.R., Lienkaemper, GW., Cromack, K.J and Cummins, K.W. 1986. Ecology of coarse woody debris in temperate ecosystem. *Adv. Ecol. Res* .**15**: 133-302.

Harrington, T.B. and Bluhm, A.A. 2001. Tree regeneration responses to microsite characteristics following a severe tornado in the Georgia Piedmont, USA. *For. Ecol. Manage.* **140**: 265-275.

Harrison, P.G and Mann, K.H. 1975. Detritus formation from eel-grass (*Zostera marina*), the relative effect of fragmentation, leaching and decay. *Limnol. Oceanogr.* **20**: 924-934.

Hasegawa, M. and Takeda, H. 1997. Carbon and nutrient dynamics in decomposing pine needle litter in relation to fungal and faunal abundances. *Pedobiologia* **40**: 171-184.

Haslam, E. 1998. *Practical Polyphenolics—From Structure to Molecular Recognition and Physiological Action.* Cambridge University Press, UK.

Haslam, S.M. 1971. Community regulation in *Phragmites communis* Trim. I. Monodominant stands. *J. Ecol.* **59**: 65-73.

Hattenschwiler, S. and Bretscher D. 2001. Isopod effects on decomposition of litter produced under elevated CO_2, N deposition and different soil types.*Global Change Biology* **7**: 565-579.

Haynes, R.J. 1986. The decomposition process: Mineralization, immobilization, humus formation and degradation. In: R.J. Haynes (Ed.), *Mineral Nitrogen in the Plant Soil System*, pp. 52-126. Academic Press. Orlando, FL.

Heady, H.F. 1956. Changes in the central California annual plant community induced by the manipulation of natural mulch. *Ecology* **37**: 798-811.

Heal, O.W. and French, D.D. 1974. Decomposition of organic matter in tundra. In: A.J. Holding, O.W. Heal, S.F. Maclean, Jr. and P.W. Flanagan. (Eds.), *Soil Organisms and Decompositions in Tundra*, pp. 227-248. IMP Tundra Biome Steering Committee, Stockholm, Sweden.

Hector, A., Beale, A.J., Minns, A., Otway, S.J. and Lawton, J.H. 2000. Consequences of the reduction of plant diversity for litter decomposition: Effects through litter quality and microenvironment. *Oikos* **90**: 357-371.

Heij, G.J. and Schneider, T. 1991. *Acidification Research in the Netherlands Studies in Environmental Science.* Vol 46. Elsevier, Amsterdam.

Hely, C., Bergeron, Y. and Flannigan, M.D. 2000. Coarse woody debris in the south-eastern Canadian boreal forest: composition and load variations in relation to stand replacement. *Can. J. For Res.* **30**: 674-687.

Henderson, S.P., Hattersley, P., von Caemmer, S. and Osmond, B. 1993. Are C_4 pathway plants threatened by global climatic change? In: E.D. Schulze and M. Cadwell (Eds.), *Ecophysiology ofPhotosynthesis*. Springer Verlag, Berlin.

Henis, Y., Tagari, H. and Volcani, R. 1964. Effect of water extracts of Carob pods, tannic acid, and their derivatives on the morphology and growth of microorganisms. *Appl. Microbiol.* **12**: 204-209.

Henssen, A. 1957a. *Beiträge zur Morphologie and Systematik der thermophilen Actinomyceten. Arch. Mikrobiol..* **26**: 373-414.

Henssen, A. 1957b. *Über die Bedeutung der thermophilen Mikroorganismen für die Zersetzung des Stallmistes. Arch. Mikrobiol.* **27**: 63-81.

Herman, R.K. and Chilcote, W.W. 1965. Effect of seedbed on germination and survival of Douglas fir. *Res. Paper Oregon For Res. Lab.* **4**: 1-28.

Herman, W.A., McGill, W.B. and Dormar, J.F. 1977. Effects of initial chemical composition on decomposition of roots of three grass species. *Canadian Journal of Soil Science* **57**: 202-215.

Hermy, M. 1987. Path analysis of standing crop and environmental variables in the field layer of two Belgian riverine forests. *Vegetatio* 70: 127-133.

Heymans, J.J., Ulanowicz, R.E. and Bondavalli, C. 2002. Network analysis of the South Florida Everglades graminoid marshes and comparison with nearby cypress ecosystems. *Ecol. Model.* **149**: 5-23.

Hirschel, G, Korner, C.H. andArnon, J.A. III. 1997. Will rising atmospheric CO_2 affect leaf litter quality and in situ decomposition rates in native plant communities? *Oecologia* **110**: 387-392.

Hobbie, S.E. 1992. Effects of plant species on nutrient cycling. *Trends in Ecology and Evolution* **7**: 336-339.

Hobbie, S.E. 2000. Interactions between litter lignin and soil nitrogen availability during leaf litter decomposition in a Hawaiian Montane forest. *Ecosystems* **3**: 484-494.

Hobbie, S.E. and Chapin HI, F.S. 1996. Winter regulation of tundra litter carbon and nitrogen dynamics. *Biogeochemisny* **35**: 327-338.

Hobbie, S.E. and Vitousek, P.M. 2000. Nutrient regulations of decomposition in Hawaiian Montane forests. Do the same nutrients limit production and decomposition? *Ecology* **81**: 1867-1877.

Hoch, W.A., Zeldin, E.L. and McCown, B.H. 2001. Physiological significance of anthocyanins during autumnal leaf senescence. *Tree Physiology* **21**: 1-8.

Holding, A.J., Franklin D.A. and Watling, R. 1965. The microflora of peat-podzol transitions. *Journal of Soil Science* **16**: 44-59.

Holland, E.A. and Coleman, D.C. 1987. Litter placement effects on microbial and organic matter dynamics in an agroecosystem. *Ecology* **68**: 425-433

Hollinger, D.Y. 1986. Herbivory and the cycling of nitrogen and phosphorus in isolated California oak trees. *Oecologia* **70**: 291-297.

Honeycutt, C.W., Zibilske, L.M. and Clapham, W.M. 1988. Heat units for describing carbon mineralization and predicting net nitrogen mineralization. *Soil Sci. Soc. Am. j* **52**: 1346-1350.

Hoondal, GS., Tiwari, R.P., Tewari, R., Dahiya, N., and Beg, Q.K. 2002. Microbial alkaline pectinases and their industrial applications: a review. *Applied Microbiology and Biotechnology* **59**: 409-418.

Hopkins, B. 1966. Vegetation of the Olokemeji forest Reserve, Nigeria. IV. The litter and soil, with special reference to their seasonal changes. *J. Ecol.* **54**: 687-703.

Homer, J.D., Gosz, J.R. and Cates, R.G. 1988. The role of carbon based metabolites in decomposition in terrestrial ecosystems. Amer. Nat. **132**: 869-883.

Houghton, J.T., Meira Filho, L.G., Callander, B.A., Harris, N., Kattenberg, A. and Maskell, K. 1996. *Climate Change 1995. The Science of Climate Change.* Cambridge University Press, Cambridge.

Howard, D.H and Gupta, R.K. 1971. Lysis of zoopathogenic fungi by streptomycetes. *Can J Microbiol.* **17**: 521-523.

Howard, J.A. and Howard, D.M. 1974. Microbial decomposition of tree and shrub leaf litter. *Oikos* **25**: 311-352.

Howard, J.A. and Howard, D.M. 1979. Respiration of decomposing litter in relation to temperature and moisture. *Oikos* **33**: 457-465.

Howarth, R.W., Fisher, S.G. 1976. Carbon, nitrogen, and phosphorus dynamics during leaf decay in nutrient enriched stream microecosystems. *Freshwater Biol.* **6**: 221-28.

Hu, S., Chapin, F.S.III, Firestone, M.K., Field, C.B. and Chiariello, M.R. 2001. Nitrogen limitation of microbial decomposition in a grassland under elevated CO_2. *Nature* **409**: 188-191.

Huang, P.M., 2002. Foreseeable impacts of soil mineral-organic component-microorganism interactions on society: Ecosystem health. In: A. Violante, P.M. Huang, J.M. Bollag and L. Gianfreda (Eds.), *Soil Mineral-Organic Matter-Microorganism Interactions and Ecosystem Health. Elsevier,* Boston.

Hulbert, L.C. 1969. Fire and litter effects in undisturbed bluestem prairie in Kansas. *Ecology* **50**: 874-877.

Humphreys, F.R. and Craig, F.G 1981. Effects of fire on soil chemical, structural and hydrological properties. In: A.M. Gill, R.H. Groves and I.R. Noble (Eds.), *Fire and the Australian Biota,* pp. 177-200. Aust. Acad. Science, Canberra, Australia.

Hungate, B.A., Holland, E.A., Jackson, R.B. Chapin, F.S. III, Mooney, H.A. and Field, C.B. 1997. The fate of carbon in grasslands under carbon dioxide enrichment. *Nature* **388**: 576-579.

Hunt, H.W. 1977. A simulation model for decomposition in grasslands. *Ecology* **58**: 469-484.

Hunt, H.W. 1978. A simulation model for decomposition in grasslands. In: G Innis (Ed.), *Grasslands Simulation Models,* pp. 155- 183. Springer, New York

Hurst, H.M. and Burges, N.A. 1967. Lignin and humic acids. In: A.D. Mac Laren and G.H. Paterson (Eds.), *Soil Biochemistry,* pp. 260-286. Edward Arnold, London.

Hunt, H.W., Coleman, D.C., Ingham, E.R., Ingham, R.E., Elliott, E.T., Moore, J.C., Rose, S.L., Reid, C.P.P. and Morley, C.R. 1987. The detrital food web in a shortgrass prairie. *Biol. Fertil. Soils* **3**: 57-68.

Huston, M.A and De Angelis, D.L. 1994. Competition and coexistence: The effects of resource transport and supply rates. *Am. Nat.* **144**: 954-977.

Ingham, R.E., Trofymow J.A., Ingham E.R. and Coleman D.C. 1985. Interactions of bacteria, fungi, and their nematode grazers-effects on nutrient cycling and plant growth. *Ecological Monographs* **55**: 119-140.

Irons, J.G, Oswood, M.W., Stout, R.J. and Pringle, C.M. 1994. Latitudinal patterns in leaf-litter breakdown: is temperature really important. *Freshw. Biol.* **32**: 401-411.

Ishizawa, S. and Araragi, M. 1970. Actinomycete flora of Japanese mills. *Soil. Sci. and. Plant Nutr.* **16**: 110-120.

Iversen, T.M., Thorup, J., Skriver, J. 1982. Inputs and transformation of allochthonous particulate organic matter in a headwater stream. *Holarct. Ecol.* **5**: 10-19.

Jackson, C., Foreman, C. and Sinsabaugh, R.L. 1995. Microbial enzyme activities as indicators of organic matter processing rates in a Lake Erie coastal wetland. *Freshwater Biol.* **34**: 329-342.

Jackson, R.B., Mooney, H.A. and Schulze, E.-D. 1997. A global budget for fine root biomass, surface area, and nutrient contents. *Proc. Natl. Acad. Sci. USA* **94**: 7362-7366.

Jagnow, G. 1957. Beiträge zur Ökologie der Streptomyceten. *Arch. Mikrobiol.* **26**: 175-191.

Jandl, R. and Sollins, P. 1997. Water-extractable soil carbon in relation to the belowground carbon cycle. *Biol. Fertility Soils* **25**: 196-201.

Jenkinson, D.S., Adams, D.E. and Wild, A. 1991. Model estimates of CO_2 emisions from soil in response to global warming. *Nature* **351**: 304-306.

Jenny, H., Gessel, S.F. and Bingham, F.T. 1949. Comparative study of decomposition rates of organic matter in temperate and tropical regions. *Soil Sci.* **68**: 419-432.

Jensen, V. 1971. The bacterial flora of beech leaves. In: T.F. Preece and C.H. Dickinson (Eds.), *Ecology of leaf surface microorganisms,* pp. 463-469. Academic press, London and New York.

Jesser, R.1998. *Effects ofproductivity on species diversity and trophic structure of detritus-based food webs within sediments of Wind Cave, south Dakota.* Masters Thesis, Department of Biological Sciences, University of Northern Colorado, Greeley, CO 80639.

Jeuniaux, C. 1955. Streptomyces production of exochitinase. *Cr Seanc. Soc Biol.* **149**: 1307-1308.

Jewell, W.T. 1971. Aquatic weed decay: Dissolved oxygen utilization and nitrogen and phosphorus regeneration. *J. Water Pollut. Control Fed.* **43**: 1457-1467.

Jobbagy, E.G. and Jackson, R.B. 2000. The vertical distribution of soil organic carbon and its relations to climate and vegetation. *Ecological Applications* **10**: 423-436.

Johansson, M.B. 1984. Litter decomposition rate at burned and clear-felled areas compared to closed stands. *Dep. For. Soils, Swed. Univ. Agric. Sci.,* Report **49**. Uppsala.

Johnson, D.W. 1992. Nitrogen retention in forest soils *J. Environ. Qual.* **21**: 1-12.

Jones, C.G., Lawton, J.H. and Shachak, M. 1994. Organisms as ecosystem engineers. *Oikos* **69**: 373-386.

Jones, D., Bacon, J.S.D., Farmer, V.C. and Webley, D.M. 1968. Lysis of cellwalls of *Mucor ranzannianus* Möller by a *Streptomyces* sp. *Antonie Leeuwenhoeck* **34**: 173-182.

Jongen, M., Jones, M.B., Hebeisen, T., Blum, H. and Hendrey, G. 1995. The effects of elevated CO_2 concentrations on root growth of *Lolium perenne* and *Trifolium repens* grown in FACE system. *Global Change Biology* **1**:361-371.

Jorgensen, J.R. and Hodges Jr, C.S. 1970. Microbial characteristics of a forest soil after twenty years of prescribed burning. *Mycologia* **62**: 721-726.

Joshi, S.R., Sharma, G.D. and Mishra, R.R. 1993. Microbial enzyme activities related to litter decomposition near a highway in a sub-tropical forest of northeast India. *Soil Biol. Biochem.* **25**: 1763-1770.

Kampichler, C., Kandeler, E., Bardgett, R.D., Jones, T.H. and Thompson, L.J. 1998. Impact of elevated atmospheric CO_2 concentration on soil microbial biomass and activity in a complex weedy field model. *Global Change Biology* **4**: 335-346.

Kapulinik, Y. 1996. Plant Growth Promotion by Rhizosphere Bacteria. In: Y. Waisel, U. Kafkafi and A. Eshel (Eds.), *Plant Roots The Hidden Half* 2nd Ed., pp. 769-770. Marcel Dekker, Inc., New York.

Karenlampi, L. 1971. Weight loss of leaf litter on forest soil surface in relation to weather at Kevo station, Finnish Lapland. *Rep. Kevo Subarct. Res. Stn.* **8**: 101-103.

Karlsson, C. and Orlander, G. 2000. Soil scarification shortly before a rich seed fall improves seedling establishment in seed tree stands of *Pinus sylvestris*. *Scand. J. For. Res.* **15**: 256-266.

Katterer, T., Reichstein, M., Andren, 0. and Lomander, A. 1998. Temperature dependence of organic matter decomposition: A critical review using literature data analyzed with different models. *Biol. Fertil. Soils* **27**: 258-262.

Kaushik, N.K. and Hynes, H.B.N. 1971. The fate of the dead leaves that fall into streams. *Arch. Hydrobiol.* **68**: 465-515.

Keever, C. 1973. Distribution of major forest species in Southeastern Pennsylvania. *Ecol. Monogr.* **43**: 303-327.

Kelman, A. 1967. *Sourcebook of Laboratory Exercises in Plant Pathology*. Freeman, San Francisco.

Kemp, P.R., Waldecker, D.G., Owenby, C.E., Reynolds, J.F. and Virginia, R.A. 1994. Effects of elevated CO_2 and nitrogen fertilization pretreatments on decomposition of tallgrass prairie leaf litter. *Plant and Soil* **165**: 115-127

Kenworthy, W.J., Currin, C. and Thayer, G. 1987. The abdundance, biomass and acetylene reduction activity of bacteria associated to decomposing rhizomes of two seagrasses, *Zostera marina and Thalassia testudinum. Aquatic Bot.* **27**: 97-119.

Khaziyev, F.K. 1977. Dynamics of the enzymic activity in the Chernozems of the Cis-ural region. *Soviet Soil Sci.* **1977**: 552-563.

Kilbertus, G 1980. Microhabitats in soil aggregates. Their relationship with bacterial biomass and size of prokaryotes present. *Revue 'd ' Ecologie et de Biologie du Sol* **17**: 543-557.

Killingbeck, K.T. 1986. The terminological jungle revisited: making a case for use of the term resorption. *Oikos* **46**: 263-264.

Killingbeck, K.T. 1988. Hurricane-induced modification of nitrogen and phosphorus resorption in an aspen clone: an example of diffuse disturbance. *Oecologia* **75**: 213-215.

Killingbeck, K.T. 1996. Nutrients m senesced leaves: keys to the search for potential resorption and resorption proficiency. *Ecology* **77**: 1716-1727.

Kirby, K.J. and Drake, C.M. 1993. *Dead Wood Matters: The Ecology and Conservation of Saproxylic Invertebrates in Britain.* English Nature, Peterborough.

Kirschbaum, M.U.F. 1995. The temperature.dependence of soil organic matter decomposition and the effect of global warming on soil organic carbon storage. *Soil Biol. Biochem.* **27**: 753-760.

Kiss, S., Dragan-Bularda, M. and Radulescu, D. 1972. Biological significance of the enzymes accumulated in soils. In: *Third Symposium on Soil Biology,* pp. 19-79. Rumanian National Society of Soil Science, Bucharest

Klein, D.A., Loh, T.C. and Goulding, R.L. 1971. A rapid procedure to evaluate the dehydrogenase activity of soils low in organic matter. *Soil Biol. Biochem.* **3**: 385-387.

Klopatek, J.M. 2002. Belowground carbon pools and processes in different age stands of Douglas fir. *Tree Physiol.* **22**: 197-204.

Knapp, A.K. and Seastedt, T.R. 1986. Detritus accumulation limits productivity of tall grass prairie. *Bioscience* **36**: 622-668.

Knops, J.M.H., Bradley, K.L. and Wedin, D.A. 2002. Mechanisms of plant species impacts on ecosystem nitrogen cycling. *Ecology Letters* **5**: 454-466.

Knosel, D. 1970. Pectolytic and cellulolytic activity of mesophilic, thermotolerant and thermophilic bacteria isolated from heated hay. *Zentralbl Bakteriol Parasitenkd Infektionskr Hyg.* 124: 190-194.

Koide, R.T., Xu, B. and Sharda, J. 2005. Contrasting belowground views of an ectomycorrhizal fungal community. *New Phytologist* **166**: 251-262.

Kok, C.J. and Velde, G.V.D. 1991. The influence of selected water quality parameters on the decay rate and exoenzymatic activity of detritus of *Nymphaea alba* L. floating leaf blades in laboratory experiments. *Oecologia* **88**: 311-316.

Kolattukudy, P.E. 2001. Polyesters in higher plants. *Advances in Biochemical Engineering/Biotechnology* **71**: 1-49.

Kolattukudy, P.E. and Espelie, K.E.1989. Chemistry, biochemistry, and function of suberin and associated waxes, In: J.W. Rowe (Ed.), *Natural products of woody plants. I: Chemicals extraneous to the lignocellulosic cell wall,* pp. 304-367. Springer-Verlag, Berlin.

Koo, B.J., Adriano, D.C., Bolan, N.S. and Barton, C.D. 2005. Root exudates and microorganisms. In: D. Hillel (Ed.), *Encyclopedia of Soils in the Environment,* pp. 421-428. Elsevier Ltd., Oxford, U.K.

Kononova, M.M. 1966. *Soil Organic Matter—Its Nature, its Role in Soil Formation and in Soil Fertility.* Oxford, U.K., Pergamon Press.

Kourtev, P.S., Ehrenfeld, J.G. and Huang, W.Z. 2002. Enzyme activities during litter decomposition of two exotic and two native plant species in hardwood forests of New Jersey. *Soil Biol. Biochem.* **34**: 1207-1218.

Kouyeas, V. 1964. An approach to the study of moisture relations of soil fungi. *Plant Soil* **20**: 351-364.

Kozlowski, T.T. 2002. Physiological ecology of natural regeneration of harvested and disturbed forest stands: Implications for forest management. *For. Ecol. Manage.* **158**: 195-221.

Kroehler, C.J. and Linkins, A.E. 1991. The absorption of inorganic phosphate from ^{32}P-labeled inositol hexaphosphate by *Eriophorum vaginatum* *Oecologia* **85**: 424-428.

Krull, E. Baldock, J. and Skjemstad, J. 2001. Soil texture effects on decomposition and soil carbon storage. *NEE workshop proceedings, 18th -20th April,* pp. 103-108.

Kshattriya, S., Sharma, G.D. and Mishra, R.R. 1992. Enzyme activities related to litter decomposition in forests of different age and altitude in northeast India. *Soil Biol. Biochem.* **24**: 265-270.

Kucera, C. and Kirkham, D. 1971. Soil respiration studies in tall grass prairie in Missouri. *Ecology*, **52**: 912-915.

Kuchaeva, A.G., Taptykova, S.D., Gesheva, R.L. and Krassilnikov, N.A. 1963. *Dokl. Akad. Nauk.* SSSR **148**:1400-1402.

Kujawinski, E.B., Hatcher, P.G. and Freitas, M.A. 2002a. High resolution fourier transform ion cyclotron resonance mass spectrometry of humic and fulvic acids: improvements and comparisons. *Anal. Chem.* **74**: 413-419.

Kujawinski, E.B., Freitas, M.A., Zang X., Hatcher, P.G., Green-Church, K.B. and Jones, R.B. 2002b. The application of electrospray ionization mass spectrometry to the structural characterization of natural organic matter. *Org. Geochem.* **33**:171-180.

Kumar, J.D., Sharma, G.D. and Misra, R.R. 1992. Soil microbial population numbers and enzyme activities in relation to altitude and forest degradation. *Soil Biol. Biochem.* **24**: 761-767.

Kunc, F. 1971. Decomposition of vanillin by soil microorganisms. *Folia Microbiol. Praha* **16**: 41-50.

Kuo, M.J. and Hartman, P.A. 1966. Isolation of amylolytic strains of *Thermoactinomyces vulgaris* and production of thermophilic actinomycete amylases. *J. Bact.* **92**: 723-726.

Kuprevich, V.F. and Shcherbakova, T.A. 1966. *Soil Enzymes*. Izdatel stvo nauka i Technika, Minsk (In Russian.) (English transl. by Indian National Scientific Document Centre, New Delhi, 1971).

Kurihara, Y. and Kikkawa, J. 1986. Trophic relations of decomposers. In: J. Kikkawa and D.J. Anderson (Eds.), *Community Ecology. Patterns and Processes,* pp. 127-160. Blackwell, New York.

Kusakabe, I., Yasui, I. and Kobayashi, T. 1969. Some properties of extracellular xylanase from *Streptomyces. J. Agric. Chem. Soc. Japan* **43**: 145 - 153.

Kuster, E. 1967. The actinomycetes. In: A. Burges and F. Raw (Eds.), *Soil Biology,* pp. 111-127. Academic Press, London.

Kutiel, P. and Inbar, M. 1993. Fire impacts on soil nutrients and soil erosion in a Mediterranean pin forest plantation. *Catena* **20**: 129-139.

Kutzner, H.J. 1968. Über die Bildung von Huminstoffen durch Streptomyceten. *Landwirtsch-Forsch* **21**: 48-61.

Ladd, J.N. 1972. Properties of proteolytic enzymes extracted from soil. *Soil Biol. and Biochem.* **4**: 227-237.

Ladd, J.N. 1985. Soil enzymes. In: D. Vaughan and R.E. Malcom (Eds.), *Soil Organic Matter and Biological Activity,* pp. 175-221. Kluwer Academic Publishers, The Netherlands.

Ladd, J.N. and Buttler, J.H.A. 1972. Short term assay of soil proteolytic enzymes using proteins and dipeptide derivatives as substrates. *Soil Biol. Biochem.* **4**: 19-30.

Ladd, J.N., Foster, R.C. and Skjemstad, J.O. 1993. Soil structure: Carbon and nitrogen metabolisni. *Geoderma* **56**: 401-434.

Ladd, J.N. and Paul, E.A. 1973. Changes in enzymic activity and distribution of acid soluble, amino acid nitrogen in soil during nitrogen immobilization and mineralization. *Soil Biol. Biochem.* **5**: 825-840.

Lambert, R.L., Lang, G.E., Reiners, W.A. 1980. Loss of mass and chemical change in decaying boles of a sub-alpine balsam fir forest. *Ecology* **61**: 1460-1473.

Lancaster, J., Hildrew, A.G. and Gjeriov, C. 1996. Invertebrate drift and longitudinal transport processes in streams. *Canadian Journal of Fisheries and Aquatic Sciences* **53**: 572-582.

Laskowski, R. and Berg, B. 1993. Dynamics of some mineral nutrients and heavy metals in decomposing forest litter. *Scand. J. For. Res.* **8**: 446-456.

Latter, P.M., Cragg, J.B. and Heal, O.W. 1967. Comparative studies on the microbiology of four moorland soils in the northern Pennines. *Journal of Ecology.* **55**: 445-464.

Lavelle, P., Bignell, D. and Lepage, M. 1997. Soil function in a changing world: The role of invertebrate ecosystem engineers. *European Journal of Soil Biology* **33**: 159-193.

Lavelle, P., Lattaud, C., Trigo, D. and Barois, I. 1995. Mutualism and biodiversity in soils. *Plant and Soil* **170**: 23-33.

Lawrence, K.L. and Wise D.H. 2000. Spider predation on forest-floor Collembola and evidence for indirect effects on decomposition. *Pedobiologia* **44**: 33-39.

Lechevalier, M.P. 1972. Description of a new species, *Oerskovia xanthineolytica, and emendation of Oerskovia* Prauser *et al. Int. J. Syst. Bact.***22**: 260-264.

Lee, K.E. 1985. *Earthworms. Their Ecology and Relationship with Soil and Land Use.* Academic Press, New York.

Leiros, M.C., Trasar-Cepeda, C., Seoane, S. and Gil-Sotres, F. 1999. Dependence of minerialization of soil organic matter on temperature and moisture. *Soil Biol. Biochem.* **31**: 327-335.

Li, C. 1996. The DNDC model. In: D.S. Powlson, P. Smith and J.U. Smith (Eds.), *Evaluation of Soil Organic Matter Models,* pp. 263-268. Springer, Berlin.

Likens, G.E. and Bomiann, F.H. 1995. *Biogeochemistly ofa forested ecosystem.* Springer, Heidelberg, Berlin.

Linkins, A.E., Sinsabaugh, R.L., McClaugherty, C.M. and Melillo, J.M. 1990. Cellulase activity on decomposing leaf litter in microcosms. *Plant and Soil* **123**: 17-25.

Lipson, D.A., Schmidt, S.K. and Monson, R.K. 1999. Links between microbial population dynamics and nitrogen availability in an alpine ecosystem. *Ecology* **80**: 1623-1631.

Liu, C., Westman, C.J., Berg, B. and Kutsch, W. 2004. Variation in litterfall-climate relationships between coniferous and broadleaf forests in Eurasia. *Global Ecol. Biogeogr.* **13**: 105-114.

Lloyd, J. and Taylor, J.A. 1994. On the temperature dependence of soil respiration. *Funct. Ecol.* **8**: 315-323.

Lopez-Real, J.M. and Swift, M.J. 1977. Formation of pseudosclerotia ('zone-lines') in wood decayed by Armillaira mellea and Stereum hirsutum III. Formation in relation to the composition of the gaseous atmosphere in wood. *Trans. Br. Mycol. Soc.* **68**: 321-325.

Lou, X., Zhu, Q., Lei, Z., van Dongen, J.L.J. and Meijer, E.W. 2004. Simulation of size exclusion chromatography for characterization of supramolecular complex- A theoretical study. *Journal of Chromatography* **1029**: 67-75.

Lousier, J.D. and Bamforth, S.S. 1990. Soil protozoa. In: D.L. Dindal (Ed.), *Soil Biology Guide,* pp. 97-136. John Wiley, New York.

Lousier, J.D. and Parkinson, D. 1978. Chemical element dynamics in decomposing leaf litter. *Canadian Journal of Botany* **56**: 2795-2812.

Lubal, P., IrokST, D., Fetsch, D. and Havel, J. 1998. The acidobasic and complexation properties of humic acids. Study of complexation of Czech humic acids with metal ions. *Talanta* **47**: 401-412.

Lundegardh, H. 1927. Carbon dioxide evolution of soil and crop growth. *Soil Science* **23**: 417-453.

Lutz, H.J. and Chandler, R.F. 1946. *Forest Soils.* John Wiley & Sons Inc., New York

Lynch, J.M. and Bragg, E. 1985. Microorganisms and soil aggregate stability. *Australian Journal of Soil Research* **27**: 411-423.

MacArthur, R.H. and Wilson, E.O. 1967. *The Theory of Island Biogeography.* Princeton University Press, USA.

MacCarthy, P. 2003. Humic Substances: what we know and what we don't know. Symposium on Natural Organic Matter in Soils & Water, Iowa State University, Ames, Iowa, March 22, 2003.

Macfadyen, A. 1963. The contribution of the microfauna to total soil metabolism. In: J. Doeksen and J. Van der Drift (Eds.), *Soil Organisms,* pp. 3-16. North Holland Publ. Co., Amsterdam.

MacLean, D.A. and Wein, R.W. 1978. Weight loss and nutrient changes in decomposing litter and forest floor material in New Brunswick forest stands. *Canadian Journal of Botany* **56**: 2730-2749.

MacMahon, J.A. and Wagner, F.H. 1985. The Mojave, Sonora and Chihuahuan deserts of North America. In: M. Evenari, I. Noy-Meir and D.W. Woodall (Eds.), *Hot Deserts and Shrublands,* pp. 105-202. Elsevier, New York.

Madge, D.S. 1965. Leaf fall and litter disappearance in tropical forests. *Pedobiologia* **5**: 273-288.

Mafongoya, P.L., Barak, P. and Reed, J.D. 2000. Carbon, nitrogen and phosphorus mineralization of tree leaves and manure. *Biol. Fertil. Soils* **30**: 298-305.

Makana, J.R. and Thomas, S.C. 2005. Effects of light gaps and litter removal on the seedling performance of six African timber species. *Biotropica* **37**: 227-237.

Marks, P.L. 1983. On the origin of the field plants of the northeastern United States. *Amer. Nat.* **122**: 210-228.

Mans, R.H., Roberts, R.D., Skeffmgton, R.A. and Bradshaw,A.D. 1983. Nitrogen and the development of ecosystem. In: J.A. Lee, S. McNeill and I.H. Rorison (Eds.), *Nitrogen as an Ecological Factor*, pp. 113-136. Blackwell, Oxford.

Marschner, H. 1999. Mineral Nutrition of Higher Plants. Academic Press, San Diego.

Marshall, K.C. 1971. Sorptive interactions between soil particles and microorganisms.In: A.D. McLaren and J. Skujins (Eds.), *Soil Biochemistry* Vol. 2, pp. 409-445. Marcel Dekker New York.

Mary, B., Recous, S., Darwis, D. and Robin, D. 1996. Interactions between decomposition of plant residues and nitrogen cycling in soil. *Plant and Soil* **181**: 71-82.

Mason, C.F. 1977. *Decomposition*. Camelot Press, Southampton.

Matile, P. 1997. The vacuole and cell senescence. *Advances* in Botanical *Research* **25**: 87-112.

Matile, P. 2000. Biochemistry of Indian summer: Physiology of autumnal leaf coloration. *Experimental Gerontology* **35**: 145-158.

Matile, P., Hortensteiner, S. and Thomas, H. 1999. Chlorophyll degradation. *Annual Review of Plant Physiology and Plant Molecular Biology* **50**: 67-95.

Mayack, D.T., Thorp, J.H. and Cothran, M. 1989. Effects of burial and floodplain retention on stream processing of allochthonous litter. *Oikos* **54**: 378-388.

McCarthy, B.C. and Facelli, J.M. 1990. Microdisturbances in oldfields and forests: Implications for woody seedling establishment. *Oikos* **58**: 27-33.

McCarty, GW., Siddaramappa, R., Wright, R.J., Codling, E.E. and Gao, G 1994. Evaluation of coal combustion byproducts as soil liming materials-their influence on soil pH and enzyme activities. *Biology and Fertility of Soils* **17**: 167-172.

McClaugherty, C.A., Aber, J.D. and Melillo, J.M. 1984. Decomposition dynamics of fine roots in forested ecosystems. *Oikos* **42**: 378-386.

McGuire, A.D., Melillo, J.M. and Joyce, L.A. 1995. The role of nitrogen in the response of forest net primary production to elevated atmospheric carbon dioxide. *Annu. Rev. Ecol. Syst.* **26**: 473-503.

McLaren. A.D. 1960. Enzyme activity in structurally restricted systems. *Enzymologia* 21: 356-364.

McLaren, A.D. 1962. Use of mole fractions in enzyme kinetics. *Arch..Biochem. Biophys.* **97**: 1-16.

McLaren, A.D. and Skujins, J.J. 1963. Nitrification by Nitrobacter agilis on surface and in soil with respect to hydrogen ion concentration. *Can. J. Microbiol* **9**: 729-731.

McLaughlin, M.J., Alston, A.M. and Martin, J.K. 1988a. Phosphorus cycling in wheat-pasture rotations. II. The role of the microbial biomass in phosphorus cycling. *Aust. J. Soil Res.* **26**: 333-342.

McLaughlin, M.J., Alston, A.M. and Martin, J.K. 1988b. Phosphorus cycling in wheat-pasture rotations. III. Organic phosphorus turnover and phosphorus cycling. Aust. J. Soil Res. **26**: 342-352.

McLay, C. 1970. A theory concerning the distance traveled by animals entering the drift of a stream. *Journal of the Fisheries Research Board of Canada* **27**: 359-370.

Means, J.E., Cromack, K. Jr., MacMillan, P.C. 1985. Comparison of decomposition models using wood density of Douglas-fir logs. *Canadian* J. *Forest Res.* **15**: 1092-1098.

Medwecka-Kornas, A. 1971. Plant litter. In: J. Phillipson (Ed.), *Methods of Study in Quantitative Soil Ecology,* pp. 24-33. Blackwell, Oxford.

Meentemeyer, V. 1978. Macroclimate and lignin control of litter decomposition rates. *Ecology* **59**: 465-472.

Meentemeyer, V., Box, E.O. and Thompson, R. 1982. World patterns and amounts of terrestrial plant litter production. Bioscience **32**: 125-128.

Meiklejohn, J. 1955. The effects of bush burning on the microflora of Kenyan upland soil. *Journal of Soil Science* **6**: 111-118.

Melillo, J.M., Aber, J.D. and Muratore, J.F. 1982. Nitrogen and lignin control of hardwood leaf litter decomposition dynamics. Ecology **63**: 621-626.

Melillo, J.M., Prentice, I.C., Farquhar, G.D., Schulze, E.D and Sala, O.E. 1996. Terrestrial biotic responses to environmental change and feedbacks to climate. In: J.T. Houghton, L.G. Filho, B.A. Callander, N. Harris, A. Kattenberg and K. Maskell (Eds.), *Climate Change 1995. The Science of Climate Change,* pp. 447-481. Cambridge University Press, Cambridge

Melin, E. 1930. Biological composition of some types of litter from North American forests. *Ecology* **11**: 72-101.

Melkania, N.P. and Singh, J.S. 1989. Ecology of Indian grasslands. In: J.S. Singh and B. Gopal (Eds.), *Perspectives in Ecology,* pp. 67-103. Jagmander Book Agency, New Delhi.

Melillo, J.M., Aber, J.D., Linkins, A.E., Ricca, A., Fry, B. and Nadelhoffer K.J. 1989. Carbon and nitrogen dynamics along the decay continuum: plant litter to soil organic matter. *Plant* and *Soil* **115**: 189-198.

Mellinger, M.V. and Mc Naughton, S.J. 1975. Structure and function of successional vascular plant communities in Central New York. *Ecol. Monogr.* **45**: 161-182.

Meyer, J.L. and O'Hop, J. 1983. Leaf shredding insects as a source of dissolved organic carbon in headwater streams. *Am. Midi. Nat.* **109**: 175-83.

Miller, J. and Georgian, T. 1992. Estimation of fine particulate transport in streams using pollen as a seston analog. *Journal of the North American Benthological Society* **11**: 172-180.

Miller, R.D. and Johnson, D.D. 1964. The effect of soil moisture tension on Carbon dioxide evolution, nitrification and nitrogen mineralization. *Soil Sci. Soc. Amer. Proc.* **28**: 644-647.

Minderman, G 1968. Addition, decomposition and accumulation of organic matter in forests. *J. Ecol.* **56**: 355-62.

Minshall, GW. and Minshall, J.N. 1978. Further evidence on the role of chemical factors in determining the distribution of benthic invertebrates in the River Duddon. *Arch. Hydrobiol.* **83**: 324-55.

Mishra, P.C. and Pradhan, S.C. 1987. Seasonal variation in amylase, invertase, cellulase activity and carbon dioxide in a tropical protected grassland of Orissa, India, sprayed with carbaryl insecticide. *Environmental Pollution* **43**: 291-300.

Mitchell, M.J. and Parkinson, D. 1976. Fungal feeding of Oribatid mites (Acari-Cryptostigmata) in an aspen woodland soil. *Ecology* **57**: 302-312.

Mohamed, M.E.S., Zaar, A., Ebenau-Jehle, C. and Fuchs, G. 2001. Reinvestigation of a new type of aerobic benzoate metabolism in the proteobacterium *Azoarcus evansii. Journal of Bacteriology* **183**: 1899-1908.

Molina, J.A.E., Hadas, A. and Clapp, C.E. 1990. Computer simulation of nitrogen turnover in soil and priming effect. *Soil Biol. Biochem.* **22**: 349-353.

Molofsky, J. and Augspurger, C.K. 1992. The effect of leaf litter on early seedling establishment in a tropical forest. *Ecology* **73**:68-77.

Montagnini, F., Ramstad, K. and Sancho, F. 1993. Litterfall, litter decomposition and the use of mulch of four indigenous tree species in the Atlantic lowlands of Costa Rica. *Agroforestry Syst.* **23**: 39-61.

Montana, C., Ezcurra, E., Camillo, A. and Delhome, J.P. 1988. Decomposition of litter in grasslands-Arid and non arid environments. *J. Arid Environ.* **14**: 55-60.

Moore, J.C., Berlow, E.L., Coleman, D.C., de Ruiter, P.C., Dong, Q., Hastings, A., Johnson, N.C., McCann, K.S., Melville, K., Morin, J.P., Nadelhoffer, K., Rosemand, A.D., Post, D.M., Sabo, J.L., Scow, K.M., Vanni, M..J. and Wall, D.H. 2004. Detritus, trophic dynamics and biodiversity. *Ecology Letters* 7: 584-600.

Moore, J.C. and Hunt, H.W. 1988. Resource compartmentation and the stability of real ecosystems. *Nature* **333**: 261-263.

Moore, J.C., McCann, K., Setälä, H. and de Ruiter, P.C. 2003. Topdown is Bottom-up: Does predation in the rhizosphere regulate aboveground dynamics? *Ecology* **84**: 846-857.

Moore, J.C., Walter, D.W. and Hunt, H.W. 1988. Arthropod regulation of micro- and mesobiota in below-ground detrital food webs. *Annu. Rev. Entomol.* **33**: 419-439.

Moore, T.R. and Dalva, M. 2001. Some controls of the release of dissolved organic carbon by plant tissues and soils. *Soil Science* **166**: 38-47.

Moorhead, D.L. and Sinsabaugh, R.L. 2000. Simulated patterns of litter decay predict patterns of extracellular enzyme activities. *Applied Soil Ecology* **14**:71-79.

Moro, M.J. and Domingo, F. 2000. Litter decomposition in four woody species in a Mediterranean climate: weight loss, N and P dynamics. *Annals of Botany* **86**: 1065-1071.

Mortland, M.M. and Wolcott, A.R. 1965. Sorption of inorganic nitrogen compounds by soil materials. In: W.V. Bartholemew and F.C. Clark (Eds.), *Soil Nitrogen,* pp. 150-197. Agron. 10. Amer. Soc. Agron; Madison, Wisconsin.

Mulholland, P.J., Tank, J.L., Webster, J.R., Dodds, W.K., Hamilton, S.K., Johnson, S.L., Marti, E., McDowell, W.H., Merriam, J.L., Meyer, J.L., Peterson, B.J., Valett, H.M. and Wollheim, W.M. 2002. Can uptake lengths in streams be determined by nutrient addition experiments'?: Results from an inter-biome comparison study. J. N. *Am. Benthol. Soc.* **21**: 544-560.

Nambiar, E.K.S. and Fife, D.N. 1991. Nutrient retranslocation in temperate conifers. *Tree Physiol.* **9**: 185-207.

Narmipieri, B., Ceccanti, B., Conti, C. and Bianchi, D. 1982. Hydrolytic enzymes extracted from soil and their properties and activities. *Soil Biol. Biochem.***14**: 257-263.

Nardi, S., Condheri, G, and Dell'Agnola, G 1996. Biological Activity of Humus. In: A. Piccolo (Ed.), *Humic Substances in Terrestrial Ecosystems,* pp. 361-406. Elsevier, Amestradam.

Neff, J.C. and Asner, G.P. 2001. Dissolved organic carbon in terrestrial ecosystems: Synthesis and a model. *Ecosystems* **4**: 29-48.

Neff, J.C., Hobbie, S.E. and Vitousek, P.M. 2000. Controls over dissolved organic C, N and P fluxes and stoichiometry in tropical soils of varying nutrient availability. *Biogeochemistry* **51**: 283-302.

Neher, D.A. 1999. Soil community composition and ecosystem processes. Comparing agricultural ecosystems with natural ecosystems. *Agroforestry systems* **45**: 159-185.

Neher, D.A., Barbercheck, M.E., El-Allaf, S.M. and Anas, 0. 2003. Effects of disturbance and ecosystem on decomposition. *App. Soil Ecol.* **23**: 165-179.

Nepstad, D.C., Carvalho, C.R. de, Davidson, E.A., Jipp, P.H., Lefebvre, P.A., Negreiros, GH., Silva, E.D., da, Stone, T.A., Trumbore, S.E. and Vieira, S.1994. The role of deep roots in the carbon cycles ofAmazonian forests and pastures. *Nature* **372**: 666-669.

Nester, E.W., Roberts, C.E. and Nester, M.T. 1995. *Microbiology: A Human Perspective.* Wm. C. Brown Communications, Inc. Dubuque, Iowa.

Niklaus, P.A., Spinier, P. and Korner, C. 1998. Soil moisture dynamics of calcareous grassland under elevated CO_2. *Oecologia* **117**: 201-208.

Nilsson, M.-C., Gallet, C. and Wallstedt, A. 1998. Temporal variability of phenolics and batatasin-III in *Empetrum hermaphroditum* leaves over an eight-year period: interpretations of ecological function. *Oikos* **81**: 6-16.

Nimlos, M.R., Blanksby, S.J., Ellison, G.B. and Evans, R.J. 2003. Enhancement of 1,2-dehydration of alcohols by alkali cations and protons, A model for dehydration of carbohydrates: *Journal of Analytical and Applied Pyrolysis* **66**: 3-27.

Nitschelm, J.J., Luscher, A., Hartwig, U.A. and van Kesssel, C. 1997. Using stable isotopes to determine soil carbon input differences under ambient and elevated atmospheric CO_2 conditions. *Global Change Biology* **3**: 411-416.

Nodvin, S.C., Driscoll, C.T. and Likens, GE. 1986. Simple partitioning of anions and dissolved organic carbon in a forest soil. *Soil Sci.* **142**: 27-35.

Norby, R.J., Cotrufo, M.F., Meson, P., 0' Neill, E.G and Canadell, J.G. 2001. Elevated CO_2, litter chemistry, and decomposition: A synthesis. *Oecologia* **127**: 153-165.

Norby, R.J., Long, T.M., Hartz-Rubin, J.S. and O'Neill, E.G. 2000. Nitrogen resorption in senescing tree leaves in a warmer, CO_2-enriched atmosphere. *Plant and Soil.* **224**: 15 -29.

Norby, R. J., Wullschleger, S.D., Gunderson, C.A., Johnson, D.W. and Ceulemans, R. 1999. Tree responses to rising CO_2: Implications for the future forest. *Plant Cell Environ.* **22**: 683-714.

Nordell, K.O. and Karlsson, P.S. 1995. Resorption of nitrogen and dry matter prior to leaf abscission: Variation among individuals, sites and years in the mountain birch. *Functional Ecology* **9**: 326-333.

Noval, J.J. and Nickerson, W.J. 1959. Decomposition of native keratin by *Streptomyces fradiae. J. Bact.* **77**: 251 -263.

Noy-Meir, I. 1985. Desert ecosystem structure and function. In. M. Evanari, I. Noy-Meir and D.W. Woodall (Eds.), *Hot Deserts and Shrublands,* pp. 93-104. Elsevier, New York.

Nykvist, N. 1961. Leaching and decomposition of litter. IV. Experiments on needle litter of *Picea abies. Oikos* **12**: 264- 279.

Nykvist, N. 1963. Leaching and decomposition of water soluble organic substances from different types of leaf and needle litter. *Studia Forest Suedica* **3**: 1-31.

Nykvist, N. 1977. Skoliga atgaders inverkan pa storlek och tillganglighet av ekosystemets narings forrad. *Sver. Skogsvards forb. tidskr.* **2- 3**: 167-178.

Oades, J.M. 1988. The retention of organic matter in soils. *Biogeochemistiy* **5**: 35-70.

Oades, J.M. 1989. An introduction to organic matter in mineral soils. In: J.B. Dixon and S.B. Weed (Eds.), *Minerals in Soil Environments,* pp. 89-106. Soil Science Society of America, Madison, WI.

Oades, J.M. 1995. Recent advances in organomineral interactions: Implications for carbon cycling and soil structure. In: P.M. Hung, J. Berthelin, J.M. Ballag, W.B. McGill and A.L. Page (Eds.), *Environmental Impact of Soil Component Interactions,* Vol. 1, pp. 119-134. Lewis Publishers: Boca Raton, FL.

Odum, E.P. 1960. Organic production and turnover in oldfield succession. *Ecology* **41**: 34-49.

Odum, E.P. 1969. The strategy of ecosystem development. *Science* **164**: 262 270.

Odum, W.E., Zieman, J.C. and Heald, E.J. 1972. The importance of vascular plant detritus to estuaries. In: R.H. Chabrech (Ed.), *Coastal Marsh and Estuary Management,* pp. 91-114. Baton Rouge, La. State Univ. Press.

Ohta, Y. 1989. Complex aliphatic and alicyclic extractives. In: J.W. Rowe, (Ed.), *Natural Products of Woody Plants. I—Chemicals Extraneous to the Lignocellulosic Cell Wall,* pp. 274-299. Springer-Verlag, Berlin.

Okafor, N. 1966. Ecology of microorganisms on chitin buried in soil. *J. Gen. Microbiol.* **44**: 311-327.

Oland, K. 1963. Changes in the content of dry matter and major nutrient elements of apple foliage during senescence and abscission. *Physiologia Plantarum* **16**: 682-694.

Olson, J.S. 1963. Energy storage and the balance of production and decomposition in ecological systems. *Ecology* **44**: 323-331.

O'Neill, E.G. 1994. Responses of soil biota to elevated atmospheric carbon dioxide. *Plant and Soil* **165**: 55-65.

O'Neill, E.G. and Norby, R.J. 1996. Litter quality and decomposition rates of foliar litter produced under CO_2 enrichment. In: G.W. Koch and H.A. Mooney (Eds.), *Carbon dioxide and Terrestrial Ecosystems,* pp. 87-103. Academic Press, New York.

Omdorff, K.A and Lang, G.E. 1981. Leaf litter redistribution in a west Virginia hardwood forest. *J. Ecol.* **69**: 225-235.

Osono, T. and Takeda, H. 2004. Potassium, calcium, and magnesium dynamics during litter decomposition in a cool temperate forest. *J. For. Res.* **9**: 23-31.

Osono, T. and Takeda, H. 2005. Limit values for decomposition and convergence process of lignocellulose fraction in decomposing leaf litter of 14 tree species in a cool temperate forest. *Ecol. Res.* **20**: 51-58.

Overby, S.T and Perry, H.M. 1996. Direct effects of prescribed fire on available nitrogen and phosphorus in an Arizona Chaparral watershed. *Arid Soil Research and Rehabilitation* **10**: 347-357.

Owen, D.F. 1978. The effect of a consumer, *Phytomiza ilicis,* on seasonal leaf fall in the holly, *Ilex aquifolium. Oikos* **31**: 268-271.

Pacheco, M.L. and Havel, J. 2001. Capillary zone electrophoretic study of uranium(VI) complexation with humic acids. *J. Radioanal. Nucl Chem.* **248**: 565-570.

Paim, U. and Beckel, W.E. 1963. Seasonal oxygen and carbon dioxide content of decaying wood as a component of the micro-environment of *Orthosana brunneum* (Forster) (Coleoptera: Cerambycidae). *Can. J. Zool.* **41**: 1133-1147.

Parnas, H. 1975. Model for decomposition of organic material by microorganisms. *Soil Biol. Biochem.* **7**: 161-169.

Parton, W.J., Scurlock, J.M.O., Ojima, D.S., Gilmanov, T.G, Scholes, RJ., Schimel, D.S., Kirchner, T., Menaud, J.C., Seastedt, T., Moya, E.G., Kamnalrut, A. and Kinyamario, J.I. 1993. Observations and modeling of biomass and soil organic matter dynamics for the grassland biome world-wide. *Global Bigeochemical Cycles* **7**: 785-809.

Pastor, J., Stillwell, M.A. and Tilman, D. 1987. Little bluestem litter dynamics in Minnesota old fields. *Oecologia* **72**: 327-330.

Paul, E.A. and Clark, F.E. 1996. *Soil Microbiology and Biochemistry.* 2nd Edition Academic Press San Diego, CA.

Paul, K. 2001. Temperature and moisture effects on decomposition.. *NEE Workshop Proceedings* 18th - 20th April, pp. 95-102.

Paul, R.W. Jr., Benfield, E.F. and Cairns, J. Jr. 1983. Dynamics of leaf processing in a medium-sized river. In: T.D. Fontaine and S.M. Bartell. (Eds.), *Dynamics of Lotic Ecosystems,* pp. 403-423. Ann Arbor Press, Ann Arbor, MI.

Pearson, T.R.H., Burslem, D.F.R.P., Mullins, C.E. and Dalling, J.W. 2003. Functional significance of photoblastic germination in neotropical pioneer trees: a seed's eye view. *Functional Ecol.* **17**: 394-402.

Pena-Mendez, E.M., Havel, H. and Patoeka, J. 2005. Humic substances—compounds of still unknown structure: applications in agriculture, industry, environment, and biomedicine. *J. Appl. Biomed.* **3**: 13-24.

Penfound, W.T. 1964. Effect of denudation on the productivity of grasslands. *Ecology* **45**: 838-845.

Perez-Harguindeguy, N., Diaz, S., Cornelissen, J.H.C., Vendramini, F., Cabido, M. and Castellanos, A. 2000. Chemistry and toughness predict leaf litter decomposition rates over a wide spectrum of functional types and taxa in central Argentian. *Plant and Soil* **218**: 21-30.

Perez, J., Munoz-Dorado, J., de la Rubia, T., and Martinez, J. 2002. Biodegradation and biological treatments of cellulose, hemicellulose and lignin—An overview. *International Microbiology* **5**: 53-63.

Perfect, T.J., Cook, A.G., Critchley, B.R., Critchely, U., Moore, R.L., Russell-Smith, A., Swift, M.J. and Yeadon, R. 1978. The effects of DDT on the populations of soil organisms and the processes of decomposition in a cultivated soil in Nigeria. In: U. Lohm and T. Person (Eds.), *Soil Organisms as Components of Ecosystems.* Ecol. Bull. **25**, pp. 565-568. Swedish Natural Science Research Council, Stockholm.

Perino, J.V. and Risser, P.G. 1972. Some aspects of structure and function in Oklahoma oldfield succession. *Bull. Torrey Bot. Club* **99**: 233-239.

Perry, D.A., Molina, R. and Amaranthus, M.P. 1987. Mycorrhizae, mycorrhizospores and reforestation: Current knowledge and research needs. *Can J. For. Res.* **17**: 929-940.

Persson, H. 1978. Root dynamics in a young Scots pine stand in central Sweden. *Oikos* **30**: 508-519.

Persson, S., Malmer, N. and Wallen, B. 1987. Leaf litter fall and soil acidity during half a century of secondary succession in a temperate deciduous forest. *Vegetatio* **73**: 31-45.

Perucci, P., Scarponi, L. and Busineli, M. 1984. Enzyme activities in a clay - loam soil amended with various crop residues. *Plant and Soil.* **81**: 345-351.

Petersen, R.C. Jr. and Cummins, K.W. 1974. Leaf processing in a woodland stream. *Freshwater Biology* **4**: 345-368.

Petersen, R.C.Jr., Cummins, K.W. and Ward, G.M. 1989. Microbial and animal processing of detritus in a woodland stream. *Ecological Monographs* **59**: 21-39.

Pickett, S.T.A., Collins, S.L. and Armesto, J.J. 1987. Models, mechanisms and pathways of succession. *Bot. Rev.* **53**: 335-371.

Pickett, S.T.A. and White, P.S. 1985. *The Ecology of Natural Disturbance and Patch Dynamics.* Academic Press, Inc., San Diego, California, USA.

Piccolo, A. 2001. The supramolecular structure of humic substances. *Soil Science* **166**: 810-832.

Piccolo, A. and Mbagwu, H.S.C. 1999. Role of hydrophobic components of soil organic matter in soil aggregate stability. *Soil Science Society of America Journal* **63**: 1808- 1810.

Polis, GA., Anderson, W.B. and Holt, R.D. 1997. Towards an integration of landscape and food web ecology: the dynamics of spatially subsidized food webs. *Ann. Rev. Ecol. System.* **28**: 289 -316.

Polis, G.A. and Hurd, S.D. 1996. Linking marine and terrestrial food webs: allochthonous input from the ocean supports high secondary productivity on small islands and coastal land communities. *Am. Nat.* **147**: 396-423.

Polis, G.A. and Strong, D.R. 1996. Food web complexity and community dynamics. *Am. Nat.* **147**: 813-846.

Polunin, N.V.C. 1984. The decomposition of emergent macrophytes in fresh water. *Adv. Ecol. Res.* **14**: 115-166.

Ponge, J.F. 1991. Food resources and diets of soil animals in a small area of Scots pine litter. *Geoderma* **49**: 33-62.

Ponge, J.F. 2003. Humus forms in terrestrial ecosystems: a framework to biodiversity. *Soil Biol. Biochem.* **35**: 935-945.

Post, W.M., Emanuel, W.R., Zinke, P.J. and Stangenberger, A.G. 1982. Soil carbon pools and world life zones. *Nature* **298**: 156-159.

Post, W.M. and Kwon, K.C. 2000. Soil carbon sequestration and land-use change: processes and potential. *Global Change Biology* **6**: 317-327.

Pragasan, L.A and Parthasarathy, N. 2005. Litter production in tropical dry evergreen forests of south India in relation to Season, plant life-forms and physiognomic groups. *Current Sci.* **88**: 1255-1263.

Pregitzer, K.S. and Euskirchen, E.S. 2004. Carbon cycling and storage in world forests: biome patterns related to forest age. *Global Change Biology* **10**: 1-26.

Prescott, C.E. 1995. Does nitrogen availability control rates of litter decomposition in forests? *Plant and Soil* **168-169**: 83-88.

Prescott, C.E., Kabzems, R. and Zabek, L.M. 1999. Effects of fertilization on decomposition rate of *Populus tremuloides* foliar litter in a boreal forest. *Canadian Journal of Forest Research* **29**: 393-397.

Pritchett, W.L. 1979. *Properties and Management of Forest Soils.* John Wiley and Sons, New York.

Proctor, J. 1984. Tropical forest litterfall II: The data set. In: A.C. Chadwick and S.L. Sutton (Eds.), *Tropical rainforest: the Leeds symposium,* Leeds Philosophical Natural History Society, Leeds.

Pugh, G.J.F. 1974. Terrestrial fungi. In: C.H. Dickinson and G.J.F. Pugh (Eds.), *Biology of Plant Litter Decomposition,* Vol. 2, pp. 303-336. Academic Press London and New York.

Pugnaire, F.I. and Chapin, F.S.III. 1993. Controls over nutrient resorption from leaves of evergreen Mediterranean species. *Ecology* **74**: 124-129.

Rabinovich, M.L., Bolobova, A.V. and Vasil'chenko, L.G. 2004. Fungal decomposition of natural aromatic structures and xenobiotics: a review. *Applied Biochem. Microbiol.* **40**: 1-17.

Raich, J.W. and Schlesinger, W.H. 1992. The global carbon dioxide flux in soil respiration and its relationship to climate. *Tellus* **44B**: 81 - 99.

Raison, R.J., Keith, H. and Khanna, P.K. 1989. Effects of fire on the nutrient - supplying capacity of forest soils. In: W.J. Dyck and C.A. Mees (Eds.), *Impact of Intensive Harvesting on Forest Site Productivity,* pp. 39-54. IEA/BE T6/ A6 Report No. 2, Forest Research Institute, Rotorua, New Zealand.

Raison, R.J., Khanna, P.K. and Woods, P.V. 1985. Mechanisms of element transfer to the atmosphere during vegetation fires. *Can. J. For. Res.* **15**: 132-140.

Rajvanshi, R. and Gupta, S.R. 1980. Decomposition of litter in a tropical dry deciduous forest. *Int. J. Ecol. Environ. Sci.* **6**: 37-49.

Ram, S.C. and Ramakrishanan, P.S. 1988. Hydrology and soil fertility of degraded grasslands at Cherrapunji in Northeastern India. *Environmental Conservation* **15**: 29-35.

Randlett, D.L., Zak, D.R., Pregitzer, K.S. and Curtis, P.S. 1996. Elevated atmospheric carbon dioxide and leaf litter chemistry: Influences on microbial respiration and net nitrogen mineralization. *Soil Sci. Soc. Am. J.* **160**: 1571-1577.

Ransom, B., Bennett, R.H. and Baerwald, R. 1997. *In situ* organic matter in recent marine sediments: A TEM investigation of organic matter preservation on continental margins and the monolayer hypothesis. *Marine Geology* **138**: 1-9.

Rastetter, E.B., Mc Kane, R.B., Shaver, G.R. and Melillo, J.M. 1992. Changes in C storage by terrestrial ecosystems: How C-N interactions restrict responses to CO_2 and temperature. *Water Air Soil Pollut.* **64**: 327-344.

Raviraja, N.S., Sridhar, K.R. and Bärlocher, F. 1998. Breakdown of *Ficus* and *Eucalyptus* leaves in an organically polluted river in India: Fungal diversity and ecological functions. *Freshwater Biol.* **39**: 537-545.

Raymond, R.L., Jamison, V.W. and Hudson, J.O. 1967. Microbial hydrocarbon co-oxidation. I. Oxidation of mono and dicyclic hydrocarbons by soil isolates of the genus *Nocardia. Appl. Microbiol.* **15**: 847-865.

Read, D.J. 1991. Mycorrhizas in ecosystems. *Experimentia* **47**: 376-391.

Reardon, M. and Forbes, A.A. 2001. Abundance and recruitment of soil invertebrates to different plant litter types in a fertigated oak plot and a control oak plot at the Falmouth sewage treatment plant.http:// courses.mbl.edu/SES/data/project/2001/reardon.pdf.

Reich, P.B. and Oleksyn, J. 2004. Global patterns of plant leaf N and P in relation to temperature and latitude. *Proc. Natl. Acad. Sci. USA* **101**: 11001-11006.

Reich, P.B., Walters, M.B. and Ellsworth, D.S. 1997. From tropics to tundra: Global convergence in plant functioning. *Proceedings of the National Academy of Sciences U.S.A.* **94**: 13730- 13734.

Reichle, D.E., McBrayer, J.F. and Ausmus, B.S. 1975. Ecological energetics of decomposer invertebrates in deciduous forest and total respiration budget. In: J. Vanek (Ed.), *Progress in Soil Zoology,* pp 283-292. Academia Publishing House of Slovak Academy of Sciences, Prague.

Reid, C.P.P. and Woods F.W. 1969. Translocation of C14-labeled compounds in mycorrhizae and its implications in interplant nutrient cycling. *Ecology* **50**: 179.

Rejmánková, E. 2005. Nutrient resorption in wetland macrophytes: comparison across several regions of different nutrient status. *New Phytologist* **167**: 471-482.

Rice, E.L. 1979. Allelopathy. An update. *Bot. Rev.* **45**: 15-109.

Rice, E.L. and Parenti, R.L. 1978. Causes of decreases in productivity in undisturbed tall grass prairie. *Amer. J. Bot.* **65**: 1091-1097.

Rich, P.H. and Wetzel, R.G 1978. Detritus in the lake ecosystem. *Am. Nat.* **112**: 57-71.

Richards, B.N. 1987. *The Microbiology of Terrestrial Ecosystems.* Longman, London.

Risley, L.S. and Crossley, D.A. 1988. Herbivore-caused greenfall in the southern Appalachians. *Ecology* **69**: 1118-1127.

Roberts, L. 1989. How fast can trees migrate? *Science* **243**: 735-737.

Robertson, G.P. and Paul. E.A. 2000. Decomposition and soil organic matter dynamics. In: O.E. Sala, R.B. Jackson, H.A. Mooney and R.W. Howarth (Eds.), *Methods in Ecosystem Science,* pp. 104-116. Springer-Verlag, New York.

Robinson, C.H., Kirkham, J.B. and Littlewood, R. 1999. Decomposition of root mixtures from high artic plants: a microcosm study. *Soil Biol. Biochem.* **31**: 1101-1108.

Rochow, J.J. 1974. Litter fall relations in a Missouri forest. *Oikos* **25**: 80-85.

Rodell, C.F. 1978. Simulation of grasshopper populations in a grassland ecosystem. In: G. Innis (Ed.), *Grassland Simulation Models,* pp. 127-154. Springer, New York.

Rodin, L.E. and Bazilevich, N.I. 1967. *Production and Mineral Cycling in Terrestrial Vegetation.* Oliver and Boyd, Edinburgh.

Rodrigo, A., Recous, S., Neel, C. and Mary, B. 1997. Modelling temperature and moisture effects on C-N transformations in soils. Comparison of nine models. *Ecological Modelling* **102**: 325-339.

Rogers, H.H., Runion. G.B. and Krupa, S.V. 1994. Plant responses to atmospheric CO_2 enrichment with emphasis on roots and the rhizosphere. *Environmental Pollution* **83**: 155-189.

Rosemond, A.D., Pringle, C.M., Ramirez, A. and Paul, M.J. 2001. A test of top-down and bottom-up control in a detritus-based food web. *Ecology* **82**: 2279-2293.

Ross, D.J. 1971. Some factors influencing the estimation of dehydrogenase activities of some soils under pasture. *Soil Biol. Biochem.* **3**: 97-110.

Ross, D.J. 1973. Some enzyme and respiratory activities of tropical soils from new herbicides. *Soil Biol. Biochem.* **5**: 559-567.

Ross, D.J. Tate, K.R., Newton, P.C.D. and Clark, H. 2002. Decomposability of C_3 and C_4 grass litter sampled under different concentrations of atmospheric carbon dioxide at natural CO_2 spring. *Plant and Soil* **240**: 275-286.

Rotundo, J.L. and Aguiar, M.R. 2005. Litter effects on plant regeneration in arid lands: a complex balance between seed retention, seed longevity and soil-seed contact. *J. Ecol.* **93**: 829-838.

Rovira, A.D. and Sands, D.C. 1971. Fluorescent pseudomonads—a residual component in the soil microflora? *J. Appl. Bact.* **34**: 253-259.

Ruddick, S.M. and Williams, S.T. 1972. Studies on the ecology of actinomycetes in soil. V. Some factors influencing the dispersal and adsorption of spores in soil. *Soil Biol. Biochem.***4**: 93-103.

Runion, GB., Curl, E.A., Rogers, H.H. Backman, P.A., Rodriguezeabana, R. and Helms, B.E. 1994. Effects of free air CO_2 enrichment on microbial populations in the rhizosphere and phylosphere of cotton. *Agricultural and Forest Meteorology* **70**: 117-130.

Rustad, L.E. and Cronan, C.S. 1988. Element loss and retention during litter decay in a red spruce stand in Maine. *Can. J. Forest Res.* **18**: 947-953.

Rypstra, A.L., Carter P.E., Balfour R.A. and Marshall S.D. 1999. Architectural features of agricultural habitats and their impact on the spider inhabitants. *Journal of Arachnology* **27**: 371-377.

Saake, B., Argyropoulos, D.S., Beinhoff, O. and Faix, O. 1996. A comparison of lignin polymer models (DHPs) and lignins by 31P-NMR spectroscopy *Phytochemistry* **43**: 499-507.

Sala, O.E. and Lauenroth, W.K. 1983. Small rainfall events: An ecological role in semiarid regions. *Oecologia* **53**: 301-303.

Salamanca, E.F., Kaneko, N., Katagiri, S. and Nagayama, Y. 1998. Nutrient dynamics and lignocellulose degradation in decomposing Quercus *serrata* leaf litter. *Ecol. Res.* **13**: 199-210.

Salas, A.M., Elliott, E.T., Westfall, D.G, Cole, C.V. and Six, J. 2003. The role of particulate organic matter in phosphorus cycling. *Soil Sci. Soc. Am. J.* **67**: 181-189.

Salfeld, J.C. and Sochtig, H. 1977. Jahreszeitliche Anderung der Humus zusammenset Zung in einem Ackerboden. *Mitt. Dtsch Boden kundl. Ges.* **25**: 265-270.

Samuelsson, J., Gustafsson, L. and Ingelog, T. 1994. *Dying and Dead Trees: A Review of Their Importance for Biodiversity.* Swedish Threatened Species Unit, Uppsala, Sweden.

Satchell, J.E. 1955. Some aspects of earthworm ecology. In: D.K. Kevan (Ed.), *Soil Zoology,* pp. 180-201. Butterworths, London.

Satchell, J.E. 1967. Lumbricidae. In: N.A. Burges and F. Row (Eds.), *Soil Biology,* pp. 259-322. Academic Press, London and New York.

Satchell, J.E. 1974. Litter interface of animate/ inanimate matter. In: C.H. Dickinson and G.J.F. Pugh (Eds.), *Biology of Plant Litter Decomposition,* Vol. 1, pp. xiii-xliv. Academic Press London, New York.

Saunders, G.W. 1976. Decomposition in freshwater. In: J.M. Anderson and A. Macfadyen (Eds.), *The Role of Terrestrial and Aquatic Organisms in Decomposition Processes,* pp. 341-73. Blackwell, Oxford, London.

Schaus, M.H. and Vanni, M.J. 2000. Effects of gizzard shad on phytoplankton and nutrient dynamics: Role of sediment feeding and fish size. Ecology **81**: 1701-1719.

Schenk, V., Manderscheid, R., Hugen, J. and Weigel, H.J. 1995. Effects of CO_2 enrichment and intraspecific ompetition on biomas partitioning, nitrogen content and microbial biomass carbon in soil of perennial ryegrass on white clover. *Journal of Experimental Botany* **46**: 987-993.

Schimel, D.S., Braswell, B.H., Holland, E.A., McKeown, R., Ojima, D.S., Painter, T. H., Parton, W.J. and Townsend, A.R. 1994. Climatic, edaphic and biotic controls over storage and turnover of carbon in soils. *Global Biogeochem. Cycles* **8**: 279-293.

Schimel, J.P. 2001. Biogeochemical models: Implicit vs. explicit microbiology. In: E.D. Schulze, S.P. Harrison, M. Heimann, E.A. Holland, J.J. Leoyd, I.C. Prentice and D. Schimel (Eds.), *Global Biogeochemical Cycles in the Climate System,* pp. 177-183. Academic press, San Diego, CA.

Schimel, J.P. and Weintraub, M.N. 2003. The implication of exoenzyme activity on microbial carbon and nitrogen limitation in soil: A theoretical model. *Soil Biol. Biochem.* **35**: 549-563.

Schlesinger, W.H. 1977. Carbon balance in terrestrial detritus. *Annual Review of Ecology and Systematics* **8**: 51-81.

Schlesinger, W.H. 1985. Decomposition of chaparral shrub foliage. *Ecology* **66**: 1353-1359.

Schlesinger, W.H. and Hasey, M.M. 1981. Decomposition of chaparral shrub foliage: Losses of organic and inorganic constituents from deciduous and evergreen leaves. *Ecology* **62**: 762-774.

Schlesinger, W.M. 1990. Evidence from chronosequence studies for a low carbon-storage potential of soils. *Nature* **348**: 232-234.

Schomberg, H.H. and Steiner, J.L. 1999. Nutrient dynamics of crop residues decomposition on a fallow no-till soil surface. *Soil Sci. Soc. Am. J.* **63**: 607-613.

Schulten, H.R. 2000. New approaches to the molecular structure and properties of soil organic matter: Humic, xenobiotic, biological, and mineral bound. In: A. Violante, P.M. Huang, J.M. Bolag and L. Gianfreda (Eds.), *Soil Mineral-Organic Matter-Microorganism Interactions and Ecosystem Health. Dynamics, Mobility and Transformations of Pollutants and Nutrients. Development in Soil Science,* pp. 351-381. Elsevier, Amsterdam.

Schulten, H.R. 2001. Models of humic structures: Association of humic acids and organic matter in soils and water. In: C.E. Clapp, M.H.B. Hayes, N. Senesi, P.R. Bloom and P.M. Jardine (Eds.), *Humic Substances and Chemical Contaminants,* pp. 26-27. Proc. Workshop and Symp. Int Humic Substances Soc., Soil Sci. Soc. Am., Am. Soc. Agronomy,USA.

Scott, D.A., Proctor, J. and Thompson, J. 1993. Ecological studies on a lowland evergreen rain forest on Maraca Island, Roraima, Brazil: II litter and nutrient cycling. *Journal of Ecology* **80**: 705-717.

Scott, N.A. and Binkley, D. 1997. Foliage litter quality and annual net N mineralization: comparison across North American forest sites. Oecologia 111: 151-159.

Seastedt, T.R. 1984. The role of microarthropods in decomposition and mineralization processes. *Annual Review of Entomology* **29**: 25-46.

Seastedt, T.R. and Crosley, D.A. 1983. Nutrients in forest litter treated with naphthalene and simulated through fall: A field microcosm study. *Soil Biol. Biochem.* **15**: 159- 165.

Semmartin, M., Aguiar, M.R., Distel, R.A., Moretto, A.S. and Ghersa, C.M. 2004. Litter. quality and nutrient cycling affected by grazing-induced species replacements along a precipitation gradient. *Oikos* **107**: 148-160.

Senthilkumar, K. 1995. *Effect of Surface Fire on the Dynamics of Beneficial Microbial Population in Natural Grassland Ecosystem, Southern India.* Ph. D thesis, Bharathiar University Coimbator, India.

Senthilkumar, K., Mani, S. and Udaiyan, K. 1997. The effect of burning on soil enzyme activities in natural grasslands in Southern India. *Ecological Research* 12: 21-25.

Senthilkumar, K., Manian, S., Udaiyan, K. and Sugavanam, V. 1995. Effect of burning.on soil nutrient status and abundance of VA- mycorrhizal fungi in a Savannah type grassland ecosystem in South India. *Tropics* **4**: 173-186.

Setala, H. and Huhta, V. 1991. Soil fauna increase *Betula pendula* growth: laboratory experiments with coniferous forest floor. *Ecology* **72**: 665-671.

Severson, K.E. and Boldt, C.E. 1978. Cattle, wildlife and rivarian habitats in the western Dakotas. In: *Management and Use of Northern Plains Rangeland. Regional Rangeland Symposium, February 27th - 28th, pp. 90-103. Prismarck, North Dakota.*

Shafizadeh, F. 1982. Introduction to pyrolysis of biomass. *Journal of Analytical and Applied Pyrolysis* **3**: 283-305.

Shafizadeh, F. 1984. The chemistry of pyrolysis and combustion. In: R. R. Roger (Ed.), *The Chemistry of Solid Wood,* pp. 489-529. American Chemical Society Advances in Chemistry Series 207 Washington.

Shaver, G.R. and Melillo, J. 1984. Nutrient budgets of marsh plants: efficiency concepts and relation to availability. *Ecology* **65**: 1491-1510.

Short, R.A., Smith, S.L., Guthrie, D.W., Stanford, J.A. 1984. Leaf litter processing rates in four Texas streams. *J. Freshwater Ecol.* **2**: 469-73.

Shure, D.J. and Gottschalk, M.R. 1985. Litter fall patterns in a floodplain forest. *Amer. Midl. Nat.* **114**: 98-111.

Silver, W.L. and Miya, R.K. 2001. Global patterns in root decomposition: comparisons of climate and litter quality effects. *Oecologia* **129**: 407-419.

Silver, W.L., Neff, J., McGroddy, M., Veldkamp, E., Keller, M. and Cosme, R. 2000. Effects of soil texture on belowground carbon and nutrient storage in a lowland Amazonian forest ecosystem. *Ecosystems* **3**: 193-209.

Singh, J.S. and Gupta, S.R. 1977. Plant decomposition and soil respiration in terrestrial ecosystems. *Bot. Review* **43**: 449-528.

Sinsabaugh, R.L. 1994. Enzymic analysis of microbial pattern and process. *Biology and Fertility of Soils* **17**: 69-74.

Sinsabaugh, R.L., Antibus, R.K. and Linkins, A.E. 1991. An enzymic approach to the analysis of microbial activity during plant litter decomposition. *Agri. Ecosys. Environ.* **34**: 43-54.

Sinsabaugh, R.L., Antibus, R.K., Linkins, A.E. McClaugherty, C.A., Rayburn, L., Repert, D. and Weiland, T. 1992. Wood decomposition over a first order watershed mass loss as a function of lignocellulose activity. *Soil Biol. Biochem.* **24**: 743-749.

Sinsabaugh, R.L, Benefield, E. and Linkins, A. III. 1981. Cellulose activity associated with the decomposition of leaf litter in a woodland stream. *Oikos* **31**: 184-190.

Sinsabaugh, R.L., Carreiro, M.M. and Alvarez, S. 2002. Enzyme and microbial dynamics , of litter decomposition. In: Richard G. Burns and Richard P. Dick (Eds.), *Enzymes in the Environment: Activity, Ecology and Applications,* pp. 249-266. Marcel Dekker, Inc. Madison Avenue, New York, USA.

Sinsabaugh, R. and Linkins, A.E. 1987. Inhibition of the *Trichoderma viridae* cellulose complex by leaf litter extracts. *Soil Biol. Biochem.* **19**: 719-725.

Sinsabaugh, R.L. and Moorhead, D.L. 1994. Resource allocation to extracellular enzyme production: A model for nitrogen and phosphorus control of litter decomposition. *Soil Biol. Biochem.* **26**: 1305-1311.

Siu, R.G.H. 1951. *Microbial Decomposition of Cellulose.* Reinhold, New York

Sjors, H. 1959. Changes in pH of leaf litter during a field experiment. *Oikos.* **10**: 225-232.

Skujins, J. 1967. Enzymes in soil. In: A.D. Maclaren and G.H. Peterson (Eds.), *Soil Biochemistry*, Vol. 1, pp. 371-414. Marcel Dekker, New York.

Skujins, J. 1973. Dehydrogenase: An indicator of biological activities in arid *soils. Bull. Ecol. Res. Commun. NFR* **17**: 235-241.

Skujins, J. 1976. Extracellular enzymes in soil. CRC *Crit. Rev. Microbiol.* **4**: 383-421.

Skujins, J.J., Potgieter, H.J. and Alexander, M. 1965. Dissolution of fungal cell walls by a streptomycete chitase and β (1-3) glucanase. *Arh. Biochem. Biophys.* **II**: 358-364.

Slovic, S. 1997. Tree physiology. In: R.F. Huttl and W. Schaaf (Eds.), *Magnesium deficiency in forest ecosystems,* pp. 101-214. Kluwer Academic, Dordrecht.

Small, J.A., Buell, M.F., Buell, H.F. and Siccama, T.G 1971. Old- field succession on the New Jersey Piedmont. The first year. *William L. Hutcheson Mem. For. Bull.* **2**: 26-30.

Smith, M.D. 1995. Forest stand regeneration: national and artificial. In: W.A. Nierenberg (Ed.), *Encyclopedia of Environmental Biology*. Vol. 2. Academic Press, San Diego, California.

Smith, V.C. and Bradford, M.A. 2003. Litter quality impacts on grassland litter decomposition are differently dependent on soil fauna across time. *Applied Soil Ecology* **24**: 197-203.

Somartne, S. and Dhanapala, A.H. 1996. Potential impact of global climate change on forest distribution in Sri Lanka. *Water Air and Soil Pollution* **92**: 129-135.

Sookne, A.M. and Harris, M. 1954. Base exchange properties. In: E. Ott, H.M. Spurlin and M.W. Graffin (Eds.), *Cellulose and Cellulose Derivatives,* Part I, pp. 208-215. Interscience, New York and London.

Sowerby, A., Blum, H., Gray, T.R.G. and Ball, A.S. 2000. The decomposition of *Lolium perenne* in soils exposed to elevated CO_2: Comparisons of mass loss of litter with soil respiration and soil microbial biomass. *Soil Biol. Biochem.* **32**: 1359-1366.

Speaker, R., Moore, K. and Gregory S. 1984. Analysis of the process of retention of organic matter in stream ecosystems. *Verhandlungen der Internationale Vereinigung fur Theoretische and Angewandte Limnologie* **22**: 1835-1841.

Speaker, R.W., Luchessa, K.J., Franklin, J.F. and Gregory, S.V. 1988. The use of plastic strips to measure leaf retention by riparian vegetation in a coastal Oregon stream. *American Midland Naturalist* **120**: 22-31.

Speir, T.W. 1977. Studies on a climosequence of soils in tussock grasslands. 10. Distribution of urease, phosphatase and sulphatase activities in soil. *New Zealand J. Sci.* **20**: 151-157.

Speir, T.W., Lee, R., Pansier, E.A. and Cairns, A. 1980. A comparison of sulfatase, urease and protease activities in planted and in fallow soils. *Soil Biol. Biochem.* **12**: 281-291.

Sprugel, D. 1984. Diversity, biomass and nutrient cycling changes during stand development in wave regenerated balsam fir forest. *Ecol. Monogr.* **54**: 165-186.

Staaf, H. and Berg, B. 1982. Accumulation and release of plant nutrients in decomposing Scots pine needle litter. Long term decomposition in a Scots pine forest. II. *Canadian Journal of Botany* **60**: 1561-1568.

Stark, J.M and Firestone M.K. 1995. Mechanisms for soil moisture effects on nitrifying bacteria. *Applied Environmental Microbiology* **61**: 218-221.

Steel, E.A., Naiman R.J. and West, S.D. 1999. Use of woody debris piles by birds and small mammals in a riparian corridor. *Northwest Science* **73**: 19-26.

Stenson, A.C., Landing, W.M., Marshall, A.G. and Copper, W.T. 2002: Ionization and fragmentation of humic substances in electrospray ionization Fourier transform-ion cyclotron resonance mass spectrometry. *Anal. Chem.* **74**: 4397-4409.

Stenson, A.C., Marshall, A.G. and Copper, W.T. 2003. Exact masses and chemical formulas of individual suwannee river fulvic acids from ultrahigh resolution ESI FT-ICR mass spectra. *Anal. Chem.* **75**: 1275-1284.

Stevenson, F.J. 1994. *Humus Chemistry—Genesis, Composition, Reactions* (2nd ed.) New York, John Wiley.

Stevenson, I.L. 1959. Dehydrogenase activity in soils. *Can. J. Microbiol.* **5**: 229-235.

Stevenson, I.L. 1967. Utilization of aromatic hydrocarbons by Arthrobacter spp. *Can. J. Microbiol.* **13**: 205-211.

Stinner, B.R., Crossley, D.A. Jr., Odum, E.P. and Tood, R.L. 1984. Nutrient budget and internal cycling of N, P, K, Ca and Mg in conventional tillage, no tillage and oldfield ecosystems on the Georgia Piedmont. *Ecology* **65**: 354 -369.

Stjohn, T.V., Coleman D.C. and Reid C.P.P. 1983. Association of vesicular-arbuscular mycorrhizal hyphae with soil organic particles. *Ecology* **64**: 957-959.

Stohlgren, T.J. 1988. Litter dynamics in two Sierran mixed forests 2: nutrient release in decomposing leaf litter. *Canadian Journal of Forest Research* **18**: 1136-1144.

Stotzky, G. 1974. Activity, ecology and population dynamics of microorganisms in soil. In: A.I. Laskin and Lechevalier (Eds.), *Microbial Ecology,* pp. 57-135. CRC Press, Cleveland.

Stowe, L.S. 1979. Allelopathy and its influence on the distribution of plants in an Illinois oldfield. *J. Ecol.* **67**: 1065-1085.

Strain, B.R. and Bazzaz, F.A. 1983. Terrestrial plant communities. In: E.R. Lemon (Ed.), *CO_2 and Plants,* pp. 177-222. Westview, Boulder, Colorado, USA.

Stulen, I. and Den Hertog, J. 1993. Root growth and functioning under atmospheric CO_2 enrichment. *Vegetatio* **104**: 99-115.

Stutzenberger, F.J. 1971. Cellulose production by *Thermomonospora curvata* isolated from municipal solid waste compost. *Appl. Microb.* **22**: 147-152.

Stutzenberger, F.J., Faufinan, A.J. and Lossin, R.D. 1970. Cellulolytic activity in municipal waste composting. *Can. J. Microbiol.* **16**: 553-560.

Subba Rao, N.S. 2000. *Soil Microbiology* (Fourth Edition of Soil Microorganisms and Plant Growth) Oxford and IBH Pub., New Delhi, India.

Suberkropp, K. 1992a. Interactions with invertebrates. In: F. Barlocher (Ed.), *The Ecology of Aquatic Hyphomycetes,* pp. 118-133. Ecological Studies, Vol. 94. Springer, Berlin.

Suberkropp, K. 1992b. Aquatic hyphomycete communities. In: GC. Carroll and D.T. Wicklow (Eds.), *The Fungal Community. Its Organization and Role in Ecosystem, 2nd* edition, pp. 729-747. Marcel Dekker, New York.

Suberkropp, K. 1998. Microorganisms and organic matter decomposition. In: R.J. Naiman and R.E. Bilby (Eds.), *River Ecology and Management: Lessons from the Pacific Coastal Ecoregion,* pp. 120-141 Springer, New York

Suberkropp, K., Godshalk, GL. and Klug, M.J. 1976. Changes in the chemical composition of leaves during processing in a woodland stream. *Ecology* 7: 20-27.

Suberkropp, K. and Klug, M.J. 1976. Fungi and bacteria associated with leaves during processing in a woodland stream. *Ecology* **57**: 707-719.

Suberkropp, K. and Klug, M.J. 1980. The maceration of deciduous leaf litter by aquatic hyphomycetes. *Canadian Journal of Botany* **58**: 1025-1031.

Swift, M.J., Heal, O.W. and Anderson, J.M. 1979. *Decomposition in Terrestrial Ecosystems.* Blackwell Scientific, Oxford, U.K.

Sydes, C. and Grime, J.P. 1981a. Effect of tree leaf litter on herbaceous vegetation in the deciduous woodlands. I. Field investigation. *J. Ecol.* **69**: 237-248.

Sydes, C. and Grime, J.P. 198 lb. Effects of tree leaf litter on herbaceous vegetation in the deciduous woodlands. II. An experimental investigation. *J. Ecol.* **69**: 249-262.

Szczeponska, W. 1977. The effects of remains of halophytes on the growth of *Phragmites communis* Trin. and *Typha latifolia L. Ekol. Polska* **25**: 437-446.

Tabak, H.H. and Cooke, W.B. 1968. The effects of gaseous environment on the growth and metabolism of fungi. *Bot. Rev.* **34**: 126-252.

Tabatabai, M.A. 1977. Effects of trace elements on urease activity in soils. *Soil Biol. Biochem.* **9**: 9-13.

Tabatabai, M.A. 1994. Soil enzymes In: A.L. Page, R.H. Miller and D.R. Keney (Eds.), *Methods of Soil Analysis,* pp. 775-833. American Society of Agronomy Madison, Wisconsin.

Tan, K.H. 1986. Degradation of Soil Minerals by Organic Acids. In: P.M. Huang and M. Schnitzer, (Eds.), *Interactions of Soil Minerals with Natural Organics and Microbes.* Soil Science Society of America, Special Publication No. 17, Madison, Wisconsin.

Tan, K.H., 2003. *Humic Matter in Soil and the Environment.* Marcel Dekker, New York.

Tanaka, Y. 1991. Microbial decomposition of reed *(Phragmites communis)* leaves in a saline lake. *Hydrobiologia* **220**: 119-129.

Tanaka, Y. 1993. Activities and properties of cellulase and xylanase associated with Phragmites leaf litter in a seawater lake. *Hydrobiologia* **262**: 65-75.

Tank, J.L., Webster, J.R., Benfield, E.F. and Sinsabaugh, R.L. 1998. Effect of leaf litter exclusion on microbial enzyme activity associated with wood biofilms in streams. *North Am. Benthological Soc.* **17**: 95-103.

Tao, D.L., Xu, Z.B. and Li, X. 1987. Effect of litter layer on natural regeneration of companion tree species in the Korean pine forest. *Env. Exper. Bot.* **27**: 53-66.

Tate, R.L.III 1987. *Soil Organic Matter: Biological and Ecological Effects.* John Wiley and Sons.

Tate, R.L.III and Terry, R.E. 1980a. Variations in microbial activity in histosols and its relationship to soil moisture. Appl. Environ. Microbial. **40**: 313-317.

Tate, R.L.III and Terry, R.E. 1980b. Effect of sewage effluent on microbial activities and coliform populations of Pahokee muck. J. *Environ Qual.* 9: 673-677.

Taylor, B.R. Parkinson, D. and Parsons, W.F.J. 1989. Nitrogen and lignin content as predictors of litter decay rates: A microcosm test. *Ecology* **70**: 97-104.

Thomas, H. 1997. Chlorophyll: a symptom and a regulator of plastid development. *New Phytologist* **136**: 163-181.

Thomas, H., Ougham, H., Canter, P. and Donnison, I. 2002. What stay-green mutants tell us about nitrogen remobilization in leaf senescence? *Journal of Experimental Botany* **53**: 801 -808.

Thompson, K., Grime, J.P. and Mason, G 1977. Seed germination in response to diurnal fluctuations of temperature. *Nature* **267**: 147-149.

Tian, G, B.T. Kang, and L. Brussard. 1992. Biological effects plant residues with contrasting chemical composition under humid tropical conditions—Decomposition and nutrient release. *Soil Biol. Biochem.* **24**:1051-1060.

Tiessen, H., Cuevas, E. and Chacon, P. 1994. The role of soil organic matter in sustaining soil fertility. *Nature* **371**: 783-785.

Torbert, H.A., Prior, S.A., Rogers, H.H. and Runion, GB. 1998. Crop residue decomposition as affected by growth under elevated atmospheric CO_2. *Soil Sci.***163**: 412-419.

Triska, F.J. 1970. *Seasonal distribution of aquatic hyphomycetes in relation to the disappearance of leaf litter from a woodland stream.* Ph.D thesis. Univ. Pitt., Pittsburg.

Trofymow, J.A. and Coleman, D.C. 1982. The role of bacterivorous and fungivorous nematodes in cellulose and chitin decomposition in the context of a root/ rhizosphere/ soil conceptual model. In: D.W. Freckman (Ed.), *Nematodes in Soil Ecosystem,* pp. 117-138. University of Texas, Austin.

Trotter, E.H. 1990. Woody debris, forest-stream succession, and catchment geomorphology. *Journal of the North American Benthological Society* **9**: 141-156.

Trumbore, S.E. and Harden, J.W. 1997. Accumulation and turnover of carbon in organic and mineral soils of the BOREAS northern study area. *Journal of Geophysical Research* **102**: 28817-28830.

Tu, C.M. 1982. Physical treatment and reinoculation of soil: Effects on microorganisms and enzyme activities. *Soil Biol. Biochem.* **14**: 57- 61.

Tyub, S. 2004. *Litter Accumulation and Decomposition in a Temperate Grassland of Kashmir Himalaya.* M.Phil. dissertation submitted to the University of Kashmir, Srinagar, India.

Uetz, G.W. 1979. Influence of variation in litter habitats on spider communities. *Oecologia* **40**: 29-42.

Umrit, G, and Friesen, D.K. 1994. The effect of CP ratio of plant residues added to soils of contrasting phosphate sorption capacities on P uptake by *Panicum maximum* (Jaq.). *Plant and Soil* **158**: 275-285.

van Brummelen, J. and Went, J.C. 1957. Streptosporangium isolated from forest litter in the Netherlands. *Antonie van Leeuwenhock* **23**: 385-392.

van Cleve, K., Oliver, L., Schlentner, R., Viereck, L.A. and Dyrness, C.T. 1983. Productivity and nutrient cycling in taiga forest ecosystems. *Canadian Journal of Forest Research* **13**: 747-766.

van der Valk, A.G. 1986. The impact of litter and annual plants on recruitment from the seed bank of a lacustrine wetland. *Aquatic Bot.* **24**: 13-26.

van Ginkel, J.H. and Gorissen, A. 1998. In situ decomposition of grass roots as affected by elevated atmospheric carbon dioxide. *Soil Sci. Soc. Am. J.* **62**: 951-958.

van Ginkel, J.H., Gorissen, A. and van Veen, J.A. 1996. Long term decomposition of grass roots as affected by elevated atmospheric carbon dioxide *J. Environ. Qual.* **25**: 1122-1128.

Vanni, M.J. 2002. Nutrient cycling by animals in freshwater ecosystems. *Annu. Rev. Ecol. System.* **33**: 341-370.

van Vliet, P.C.J., Beare, M.H. and Coleman, D.C. 1995. Population dynamics and functional roles of Enchytraeidae (Oligochaeta) in hardwood forest and agricultural ecosystems. *Plant and Soil* **170**: 199-207.

van't Hoff, J.H. 1898. *Lectures on Theoretical and Physical Chemistry.* Part 1. Chemical dynamics, pp. 224-229. Edward Arnold, London.

Vannote, R.L., Minshall G.W., Cummins K.W., Sedell J. R. and Cushing C. E. 1980 .River Continuum Concept. *Canadian Journal of Fisheries and Aquatic Sciences* **37**: 130-137.

Vargas, A.J. 2000. Effects of fertilizer addition and debris removal on leaf-litter spider communities at two elevations. *Journal of Arachnology* **28**: 79-89.

Vazquez-Yanes, C., Orozco-Segovia, A., Rincon, E., Sanchez-Coronado, M.E., Huante, P., Toledo, J.R. and Barradas, V.L. 1990. Light beneath the litter in a tropical forest: Effect on seed germination. *Ecology* **71**: 1952-1958.

Vazquez, F.J., Acea, M.J. and Carballas, T. 1993. Soil microbial population after wild fire. *FEMS Microbial Ecology* **13**: 93-103.

Veldkamp, H. 1955. A study of the aerobic decomposition of chitin by microorganisms. Med. Landboun., *Wagenngen* **55**: 127-174.

Verhoef, H.A. and Brussaard, L. 1990. Decomposition and nitrogen mineralization in natural and agro-ecosystems: The contributions of soil animals. *Biogeochemistry* **11**: 175-211.

Vilà, M., Vayreda, J. and Gracia, C. 2004. Biodiversity correlates with regional patterns of forest litter pools. *Oecologia* **139**: 641-646.

Viro, P.J. 1974. Effects of forest fire on soil. In: T.T. Kozolowski and C.E. Ahlgren (Eds.), *Fire and Ecosystem,* pp 7 - 45. Academic Press, London.

Visser, S., Griffiths, C. L. and Parkinson, D. 1983. Effects of surface mining on the microbiology of a prairie site inAlberta. *Can. J. Soil Sci.* **63**: 177-189.

Vitousek, P.M. 1982. Nutrient cycling and nutrient use efficiency. *Am. Nat.* **119**: 553-572.

Vitousek, P.M. 1984. Litterfall, nutrient cycling, and nutrient limitation in tropical forests. *Ecology* **65**: 285-298.

Vitousek, P.M., Aber, J.D., Howarth, R.W., Likens, GE., Matson, PA., Schindler, D.W., Schlesinger, W.H. and Tilman, D.G 1997. Human alteration of the global nitrogen cycle: sources and consequences. *Ecol. Appl.* **7**: 737-750.

Vitousek, P.M. and Sanford, R.L. 1986. Nutrient cycling in moist tropical forest. *Annu. Rev. Ecol. Syst.* **17**: 137-167.

Vogt, K.A., Grier, C.C. and Vogt, D.J. 1986. Production, turnover, and nutrient dynamics in above and below-ground detritus of world forests. *Advances. in Ecological Research* **15**: 303-377.

Waddell, K.L. 2002. Sampling coarse woody debris for multiple attributes in extensive resource inventories. *Ecological Indicators* **1**: 139-153.

Wagner, J.D., Toft, S. and Wise, D.H. 2003. Spatial stratification in litter depth by forest-floor spiders. *Journal of Arachnology* **31**: 28-39.

Wahebh, M. and Mahasheh, A.M. 1985. Some aspects of decomposition of leaf litter of the sea grass Halophila stipulacea from the gulf of Aqaba (Jordan). *Aquatic Bot.* **21**: 237-244.

Waksman, S. A. 1919. Studies in the metabolism of Actinomycetes. II. *J Bact.* **4**: 307-330.

Walker, J., Raison, R.J. and Khanna, P.K. 1986. Fire. In: J.S. Russell and R.F. Isbell (Eds.), *Australian Soils: The Human Impact,* pp. 185-216. University of Queens land Press, St Lucia, Australia.

Wall, D.H., Adams, G and Parsons, A.N. 2001. Soil biodiversity. In: F.S. Chapin III, O.E. Sala and E.H. Sannwald (Eds.), *Global Biodiversity in a Changing Environment: Scenarios for the 21st Century,* pp. 47-82. Springer-Verlag, New York.

Wallace, J.B. and Benke A.C. 1984. Quantification of wood habitat in sub-tropical coastal-plain streams. *Canadian Journal of Fisheries and Aquatic Sciences* **41**: 1643-1652.

Wallace, J.B., Eggert, S.L., Meyer J.L. and Webster J.R. 1997. Multiple trophic levels of a forest stream linked to terrestrial litter inputs. *Science* **277**: 102-104.

Wallace, J.B., Webster, J.R. and Cuffney, T.F. 1982. Stream detritus dynamics: regulation by invertebrate consumers. *Oecologia* **53**: 197-200.

Wallwork, J.A. 1976. *The Distribution and Diversity of Soil Fauna.* Academic Press, New York.

Walsh, R.P.D. and Voight, P.J. 1977. Vegetation litter: An underestimated variable in hydrology and geomorphology. *J. Biogeogr.* **4**: 253-274.

Wander, M. 2004. Soil organic matter fractions and their relevance to soil function. In: F. Magdoff and R. Weil (Eds.), *Soil organic matter in sustainable agriculture,* CRC Press, London.

Ward, GM., Ward, A.K., Dahm, C.N. and Aumen, N.G. 1994. Origin and formationof organic and inorganic particles in aquatic systems. In: R.S. Wotton (Ed.), *The Biology of particles in aquatic systems,* pp. 45-73. Lewis Boca Raton, Fl.

Wardle, D.A. 1999. How soil food webs make giants grow. *Trends in Ecology & Evolution* **14**: 418-420.

Wardle, D.A. 2002. *Communities and Ecosystems: Linking Aboveground and Belowground Components.* Princeton University Press, Princeton, NJ.

Wardle, D.A., Bonner, K.I. and Nicholson, K.S. 1997. Biodiversity and plant litter: experimental evidence which does not support the view that enhanced species richness improves ecosystem function. *Oikos* **79**: 247-258.

Wardle, D.A., Bonner, K.I. and Barker, GM. 2002. Linkages between plant litter decomposition, litter quality and vegetation responses to herbivores. *Funct. Ecol.* 16: 585-595.

Wardle, D.A. and Lavelle, P. 1997. Linkages between soil biota, plant litter quality and decomposition. In: G. Cadish and K.E. Giller (Eds.), *Driven by Nature—Plant Litter Quality and Decomposition,* pp. 107-124. CAB International, Wallingford.

Waring, R.H. and Schlesinger, W.H. 1985. *Forest Ecosystems. Concepts and Management.* Academic press, New York.

Wasilewska, L., Jakubczuk, H. and Paplinska, E. 1975. Production of *Aphelenchus avenae* Bastian (Nematoda) and reduction of mycelium of saprophytic fungi by them. *Polish Ecological Studies* **1**: 61-73.

Watt, A.S. 1974. Senescence and rejuvenation in ungrazed chalk grassland (Grassland B) in Breckland: The significance of litter and moles. *J Appl. Ecol.* **23**: 1157-1171.

Weaver, J.E. and Rowland, N.W. 1952. Effect of excessive natural mulch on the development, yield, and structure of a native grassland. *Bot. Gaz.* **114**: 1-9.

Webster, J. 1957. Sucession of fungi on decaying cocksfoot culms. (part 2). *J. Ecol.* **45**: 1-30.

Webster, J.R. 1983. The role of benthic macroinvertebrates in detritus dynamics of streams: a computer simulation. *Ecol. Monogr* **53**: 383-404.

Webster, J.R. and Benfield E.F. 1986. Vascular plant breakdown in freshwater ecosystems. *Annual Review of Ecology and Systematics.* **17**: 567-594

Webester, J.R., Benfield, E.F., Ehrman,T.P., Schaeffer, M.A., Tank, J.L., Hutchens, J.J and D' Angelo, D.J. 1999. What happens to allochthonous material that falls into streams? A synthesis of new and published information from Coweeta. *Freshwater Biology* **41**: 687-705.

Webster, J.R., Benfield, E.F., Golladay, S.W., Hill, B.H., Hornick, L.E., Kazmierczak, R.F. and Perry, W.E. 1987. Experimental studies of physical factors affecting seston transport in streams. *Limnology and Oceanography* **32**: 848-863.

Webster, J.R., Covich, A.P., Tank, J.L. and Crockett, T.V. 1994. Retention of coarse organic particles in streams in the southern Appalachian Mountains. *Journal of the North American Benthological Society.* **13**: 140-150.

Webster, J. and Dix, N.J. 1960. Succession of fungi on decaying cocksfoot culms. III. A comparison of the sporulation and growth of some primary saprophytes on stem, leaf blade and leaf sheath. *Trans. Br. Mycol. Soc.* **43**: 85-99.

Webster, J.R., Wallace, J.B. and Benfield, E.F. 1995. Organic processes in streams of the eastern United States. In: C.E. Cushing, G.W. Minshall and K.W. Cummins (Eds.), *Ecosystems ofthe World 22: River and Stream Ecosystems,* pp. 117-187. Elsevier, Amsterdam.

Weir, F.L., Park, Sang-Wook and Vivanco, J.M. 2004. Biochemical and physiological mechanisms mediated by allelochemicals. *Current Opinion in Plant Biol.* **7**: 472-479.

Welbank, P.J. 1963. Toxin production during decay of *Agropyron repens* (couch grass) and other species. *Weed Res.* **3**: 205-214.

Wershaw, R.L. 1986. A new model for humic materials and their interactions with hydrophobic chemicals in soil-water or sediment-water systems. *Journal of Contaminant Hydrology* **1**: 29-45.

Wershaw, R.L. 1994. Membrane-micelle model for humus in soils and sediments and its relation to humification. U.S. Geological Survey Water-Supply Paper 2410, US Government Printing Office, 48 pp.

Wershaw, R.L. 2004. Evaluation of conceptual models of natural organic matter (humus) from a consideration of the chemical and biochemical processes of humification. *Scientific Investigations Report* **2004-5121**: 1-49, US Geological Survey, Virginia.

Wershaw, R.L. and Kennedy, K.R. 1998. Use of 13C-NMR and FTIR for elucidation of degradation pathways during natural litter decomposition and composting. IV. Characterization of humic and fulvic acids extracted from senescent leaves. In: G. Davies, and E.A. Ghabbour (Eds.), *Humic substances- Structures, properties and uses,* Cambridge, pp. 60-68. Royal Society of Chemistry, London.

West, N.E. 1979. Formation, distribution and function of plant litter in desert ecosystems. In: J.A. Perry and D.W. Goodall (Eds.), *Arid Land Ecosystems. Structure, Function and Management,* Vol. 16, pp. 647-659. Cambridge University Press, Cambridge.

Wetzel, R.G. 1995. Death, detritus, and energy-flow in aquatic ecosystems. *Freshwater Biol.* **33**: 83-89.

Weyers, H.S. and Suberkropp, K. 1996. Fungal and bacterial production during the breakdown of yellow poplar leaves in two streams. *Journal of the North American Benthological Society* **15**: 408-420.

Whitford, W.G., Freckman, D.W., Santos, P.F., Elkins, N.Z. and Parker, L.W. 1982a. The role of nematodes in decomposition in desert ecosystems. In: D.W. Frecicman (Ed.), *Nematodes in Soil Ecosystems,* pp. 98-115. University of Texas, Austin.

Whitford, W.G, Meentemeyer, V., Seastedt, T.R., Comack Jr. K., Crossley, D.A., Santos, P., Todd, R.L. and Waide, J.B. 1982b. Exceptions to the AET model: Deserts and clearcut forest. *Ecology* 62: 275-277.

Whittaker, R.H. 1970. *Communities and ecosystems.* MacMillan, New York.

Whittaker, R.H. and Woodwell, GM. 1969. Structure, production, and diversity of the oak-pine forest at Brookhaven. *J. Ecol.* **57**: 157-174.

Wieder, R.K., Lang, G.E. 1982. A critique of the analytical methods used in examining decomposition data obtained from litter bags. *Ecology* **63**: 1636-1642.

Wiegert, R.G, Coleman, D.0 and Odum, E.P. 1970. Energetics of the litter soil subsystem. In: J. Phillipson (Ed.), *Methods of Study in Soil Ecology,* pp. 93-98. IBP/UNESCO Symp., Paris.

Wiegert, R.G. and Evans, F.C. 1964. Primary production and disappearance of dead vegetation on an oldfield in south Eastern Michigan. *Ecology* **45**: 49-63.

Williams, R.J. and Ashton, D.H. 1987. Effect of disturbance and grazing by cattle on the dynamics of heathlands and grasslands communities on the Bogong High Plains, Victoria. *Austral. J. Bot.* **35**: 416-431.

Williams, S.T., Davies, F.L., Mayfield, C.I. and Khan, M.R. 1971. Studies on the ecology of actinomycetes in soil. II. The pH requirements of streptomycetes from two acid soils. *Soil Biol. Biochem.* **3**: 187-195.

Winn, A.A. 1985. Effects of seed size and microsite on seedling emergence of *Prunella vulgaris* in four habitats. *J. Ecol.* **73**: 831-840.

Wise, D.H. 1995. *Spiders in Ecological Webs.* Cambridge University Press, Cambridge, UK.

Wise, D.H., Snyder, W.E., Tuntibunpakul, P. and Halaj, J. 1999. Spiders in decomposition food webs of agroecosystems: Theory and evidence. *Journal of Arachnology* **27**: 363-370.

Witkamp, M. 1966. Decomposition of leaf litter in relation to environment, microflora and microbial respiration. *Ecology* **47**: 194-201.

Witkamp, M. and van der Drift, J. 1961. Breakdown of forest litter in relation to environmental factors. *Plant and Soil* **15**: 295-311.

Wolniewicz-Czerwinska, K.1956. Novitates Systematicae Plantorum non Vasculariun. *Acta Soc.Bot.Pol.* **25**:111-118.

Woodwell, G. 1978. The carbon dioxide question. *Sci. Am.* **238**: 34-43.

Wright, I.J. and Westoby, M. 2003. Nutrient concentration, resorption and lifespan: leaf traits of Australian sclerophyll species. *Functional Ecol.* 17: 10-19.

Wright, M.A. 1972. Factors governing ingestion by the earthworm *Lumbricus terrestris* with special reference to apple leaves. *Annals of Applied Biology* **70**: 175-188.

Xiong, S., Johansson, M.E., Hughes, F.M.R., Hayes, Richards, K. A. and Nilsson, C. 2003. Interactive effects of soil moisture, vegetation canopy, plant litter and seed addition on plant diversity in a wetland community. *J. Ecol.* **91**: 976-986.

Xiong, S. and Nilsson, C. 1999. The effects of plant litter on vegetation: a meta-analysis. *Journal of Ecology* **87**: 984-994.

Yeates, G.W. and Coleman, D.C. 1982. Nematodes in decomposition. In: D.W. Greckman (Ed.), *Nematodes in Soil Ecosystem.* pp. 55-80. University of Texas, Austin.

Zaar, A., Eisenreich, W., Bacher, A., and Fuchs, G 2001. A novel pathway of aerobic benzoate catabolism in the bacteria *Azoarcus evansii* and *Bacillus stearothermophilus. The Journal of Biological Chemistry* **276**: 24997-25004.

Zak, D.R., Pregitzer, K.S., Curtis, P.S., Teeri, J.A., Foge, R. and Randlett, D.L. 1993. Elevated atmospheric CO_2 and feedback between carbon and nitrogen cycles. *Plant and Soil* **151**: 105-117.

Zak, D.R., Pregitzer, K.S., King, J.S. and Holmes, W.E. 2000. Elevated atmospheric CO_2, fine roots and the response of soil microorganisms: A review and hypothesis. *New Phytol.* **147**: 201-22.

Zhu, W. and Ehrenfeld, J.G. 1996. The effects of mycorrhizal roots on litter decomposition, soil biota, and nutrients in a spodic soil. *Plant and Soil* **179**: 109-118.

•••

INDEX

I

L

M

N

•••

FSC
www.fsc.org
MIX
Papier | Fördert
gute Waldnutzung
FSC® C083411

Zeitfracht Medien GmbH
Ferdinand-Jühlke-Straße 7
99095 Erfurt, Deutschland
produktsicherheit@kolibri360.de